AF411817

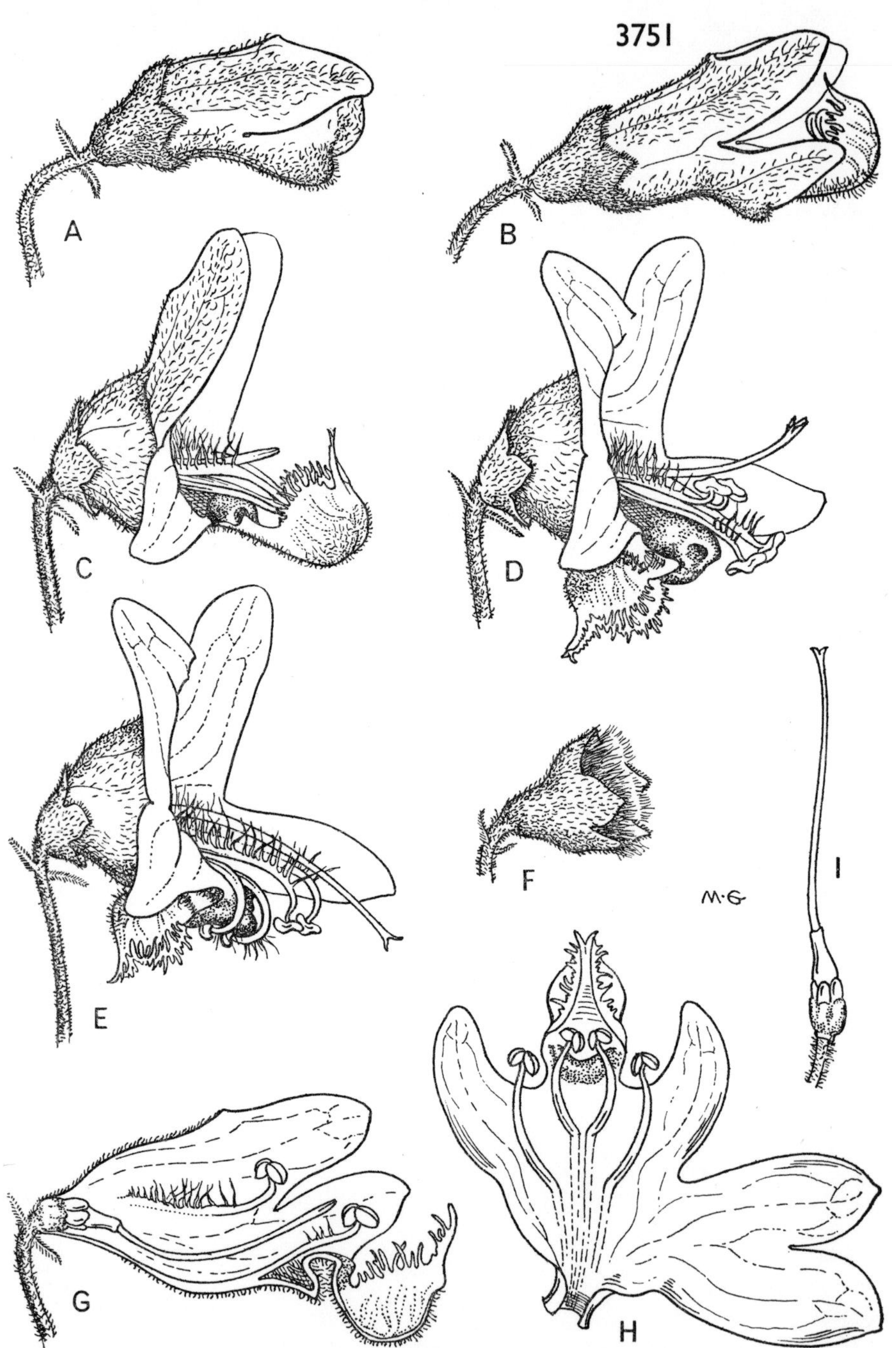

TAB. 3751. *Eriope* flower—stages in anthesis to show explosive pollination mechanism. A, flower in bud; B, flower, early stage of anthesis; C, flower, pollination mechanism untriggered; D, flower, pollination mechanism immediately after being triggered by pollinating bee, note position of lower lip and stamens, and growth of style; E, flower, late stage, note position of stamens and style; F, flower, after corolla fall, note hairs in calyx throat; G, flower vertical section, note hinge mechanism of lower lip; H, corolla, opened to display insertion of stamens, hairs on filaments not shown; I, gynoecium, note the conical stylopodium which arises between and overtops the young nutlets. (All × 6 approx., based on living material of *E. hypenioides* Mart. ex Benth.)

HOOKER'S
ICONES PLANTARUM

VOL. XXXVIII, PART III,
OF THE ENTIRE WORK

OR FIFTH SERIES VOL. VIII PART III

A REVIEW OF ERIOPE AND ERIOPIDION (LABIATAE)

R. M. HARLEY

BENTHAM-MOXON TRUSTEES
1976

Printed in Great Britain by
William Clowes & Sons Ltd
London, Colchester and Beccles

CONTENTS

ABSTRACT

The genus *Eriope* (Labiatae) consists of twenty species, all in South America, of which eighteen are restricted to Brazil. One species previously included in *Eriope* is here separated as a distinct genus *Eriopidion*.

In Part I, after a brief summary of previous taxonomic treatments, the geographical distribution of the taxa is considered in relation to their probable evolutionary history. A further section discusses the adaptive significance of the explosive pollination mechanism and other floral characters. Some of the species in the present account have been incorporated from various sections of the large neighbouring genus *Hyptis*. The evidence for this has also been used in a re-evaluation of characters in the generic and sectional delimitation between *Eriope* and related taxa in *Hyptis*. This is followed by a key to South American genera of *Hyptidinae*.

Part II contains the systematic treatment of *Eriope* and *Eriopidion*, accompanied by a key to species of the former. There is a full citation of authorities and types of accepted taxa and of those in synonymy. All material examined by the author up to the end of 1972 has been cited except for the widespread *Eriope macrostachya* var. *macrostachya* and var. *hypoleuca* and *E. crassipes* subsp. *crassipes* where a selection has been given. All taxa recognised in the account are illustrated.

The following new taxa or combinations have been made:—

Eriope latifolia (Mart. ex Benth.) Harley, comb. nov.

E. hypoleuca (Benth.) Harley, comb. nov.

E. salviifolia (Pohl ex Benth.) Harley, comb. nov.

E. exaltata Harley, sp. nov.

E. macrostachya Mart. ex Benth. var. *grandiflora* (Epling) Harley, comb. & stat. nov.; var. *platanthera* (Epling & Mathias) Harley, comb. & stat. nov.

E. parvifolia Mart. ex Benth. subsp. *glandulosa* Harley, subsp. nov.

E. tumidicaulis Harley sp. nov.

E. crassipes Benth. subsp. *trichopoda* (Briq.) Harley, comb. & stat. nov.; subsp. *cristalinae* Harley, subsp. nov.

E. arenaria Harley, sp. nov.

E. xavantium Harley, sp. nov.

Eriopidion Harley, gen. nov.

Eriopidion strictum (Benth.) Harley, comb. nov.

RESUMO

O género *Eriope* (Labiatae) é formado por vinte espécies, todas da América do Sul, 17 das quais estão circunscritas ao Brasil. Uma espécie, anteriormente incluida em *Eriope*, é agora separada constituindo um género distinto (*Eriopidion*). Na primeira parte, depois de um breve sumário sôbre anteriores trabalhos taxonómicos, considera-se a distribuição geográfica em relação à provavel história evolutiva das espécies. Seguidamente discute-se a significância adaptativa do especial mecanismo da polinização e de outros caracteres florais. Algumas das espécies incluidas neste trabalho estavam incorporadas em diversas secções do amplo e semelhante género *Hyptis*, tendo sido possivel a sua separação pela reavaliação de algums caracteres usados na delimitação genérica e das secções de *Eriope* e dos taxones semelhantes de *Hyptis*. Uma chave dicotómica para os géneros sul-americanos da Subtribo *Hyptidinae* conclui a primeira parte.

A segunda parte inclui o tratamento sistemático de *Eriope* e de *Eriopidion*, acompanhado de chaves para as espécies do primeiro género. A citação bibliográfica é o mais completa possivel, assim como a dos tipos dos diferentes taxones e respectivos sinónimos. Citam-se todos os espécimes observados pelo autor até ao fim de 1972, exceptuando-se para 3 amplos taxones (*Eriope macrostachya* var. *macrostachya* e var. *hypoleuca* e *E. crassipes* subsp. *crassipes*) nos quais se apresenta uma selecção do material estudado. É apresentada a iconografia de todos os taxones considerados nesta revisão. Os novos taxones criados e as novas combinações feitas neste trabalho, são:

Eriope latifolia (Mart. ex Benth.) Harley, comb. nov.

E. hypoleuca (Benth.) Harley, comb. nov.

E. salviifolia (Pohl ex Benth.) Harley, comb. nov.

E. exaltata Harley, sp. nov.

E. macrostachya Mart. ex Benth. var. *grandiflora* (Epling) Harley, comb. & stat. nov.; var. *platanthera* (Epling & Mathias) Harley, comb. & stat. nov.

E. parvifolia Mart. ex Benth. subsp. *glandulosa* Harley, subsp. nov.

E. tumidicaulis Harley, sp. nov.

E. crassipes Benth. subsp. *trichopoda* (Briq.) Harley, comb. & stat. nov.; subsp. *cristalinae* Harley, subsp. nov.

E. arenaria Harley, sp. nov.

E. xavantium Harley, sp. nov.

Eriopidion Harley, gen. nov.

Eriopidion strictum (Benth.) Harley, comb. nov.

PART I

Introduction

The Tribe *Ocimoideae*, to which *Eriope* and *Eriopidion* belong, are characterised by the deflexed stamens which rest on the lower lip of the corolla. The two
genera are further placed, together with *Hyptis, Marsypianthes, Peltodon* and
Raphiodon, in the Subtribe *Hyptidinae*, on account of the boat-shaped, strongly
laterally compressed lower lip which is contracted and thickened near the base
and typically encloses the anthers under tension.

The mainly Brazilian genus *Eriope* was first proposed by Kunth, but not
validly published until taken up by Bentham in 1833 in his classic work,
Labiatarum Genera et Species. At that time Bentham's classification was
largely based on the interesting new Brazilian material being provided by
von Martius, by St.-Hilaire and by Sellow. The necessity of defining his genera
and species on only a few specimens imposed severe limitations, and it is
therefore remarkable how well Bentham's classification has stood the test of
time. Further collections by Gardner, Pohl and others allowed Bentham to
extend his treatment of the genus in De Candolle's Prodromus vol. 12 (1848)
where a total of seventeen species of *Eriope* were recognised.

Originally the genus seemed a distinctive one. To quote Bentham (1833),
it was 'allied to *Hyptis*, and especially to the section *Hypenia*, but remarkably
constant both in habit and characters in all the species I have examined'.
However, in his original account he placed two species in *Hyptis* Sect. *Siagonarrhen* as *H. salviifolia* and *H. latifolia*, commenting 'species inter *Hypeniam*,
Siagonarrhen et *Eriope* quasi intermedia'. Today, *Hyptis* Sect. *Hypenia* is
still recognised though in a modified form, but Sect. *Siagonarrhen* has been
dismembered, its component species being distributed in Sect. *Hypenia*, Sect.
Buddleioides and in *Eriope*. Bentham distinguished the latter genus by its
turbinate, indistinctly and irregularly toothed calyx with its throat closed in
fruit by dense white hairs, and by the deflexed fruiting pedicel. Now it is
necessary to include other species which do not possess all these characters,
if the genus is to remain a natural one.

During the latter part of the nineteenth century John Briquet in Geneva
turned his attention to the group, describing three species of *Eriope* of which
only one, reduced to the rank of subspecies, is now included. His most significant contribution, however, was the introduction of sectional and subsectional
categories into *Eriope*.

In 1936 a major contribution on the group appeared under the title 'Synopsis
of the South American Labiatae' by Carl Epling. In this work he listed a total
of eighteen species of *Eriope*, including four new ones of which two remain
unaltered in the present account. Epling recognised two sections, Sect.
Tubiflorae containing the single species *E. stricta*, which is now made the type
of a monotypic genus, *Eriopidion*, and Sect. *Platanthera* including the remaining species of *Eriope* including the type of the genus. This name must
therefore be replaced by Sect. *Eriope*, according to the International Code

of Botanical Nomenclature. Epling further subdivided this into five subsections, partly based on Briquet's system but with substantial modifications. In the light of present knowledge these now seem unnecessarily elaborate, and rather unnatural.

Apart from a few additional species described subsequently by Epling, no further work has been done on the genus until now. The present account was stimulated firstly by my participation in the Xavantina–Cachimbo Expedition to northern Mato Grosso, sponsored by the Royal Society, the Royal Geographical Society and the National Research Council of Brazil. The greatest single contribution in recent times to a study of the Brazilian flora has undoubtedly been the fine series of collections made by Dr. Howard S. Irwin and his colleagues from the New York Botanical Garden, chiefly in the Brazilian Planalto. Thanks to the kind invitation of Dr. Irwin, I was fortunate in accompanying him in the field in 1971, and as a result have gained a much deeper knowledge of the group under review. Apart from routine field studies, the cultivation of plants from wild-collected seed, cytological examination, the collection of fresh flowers in spirit, and studies on pollination of *Eriope* and *Hyptis* have all helped to provide evidence as to relationships and possible evolutionary trends within the *Hyptidinae*.

Evolution and Variation

Eriope is centred on the seasonal upland savanna woodlands of Minas Gerais, Bahia, Goiás and neighbouring states of central Brazil, areas notable for their high endemism and floristic richness. Of the twenty species of *Eriope* recognised here, two are much more widespread. *E. crassipes* Benth. and *E. macrostachya* Mart. ex Benth. are frequent in Minas Gerais, where they both possess close relatives with very limited distribution, but their own ranges extend far into other areas of Tropical South America. *E. crassipes*, with a characteristic swollen woody rootstock and herbaceous stems, is a fairly typical member of the central Brazilian 'cerrado' formations of woodland savanna. Though most frequent in this area, it extends southwards as far as Paraná, and as subsp. *trichopoda* (Briq.) Harley, into northern Paraguay. Northwards it is an occasional component of the 'campinas' or local grassland areas which are interspersed in tropical rain forest, as in Amazonia and in French Guiana. The other widespread species, *E. macrostachya*, is a diffuse shrub most abundant in the eastern Planalto but extending from Ceará in north-east Brazil to Paraná and Paraguay in the south. Apart from this, there is a remarkably disjunct population north of the equator in Venezuela which extends the range of the genus into the Andes.

Of the species with more restricted ranges, twelve occur only in Minas Gerais and Bahia, with *E. latifolia* also from São Paulo State, and four others extend into Goiás. The remaining species *E. xavantium* Harley is a recent discovery made on the Xavantina–Cachimbo Expedition to northern Mato Grosso at the northern limits of the Brazilian Planalto.

Certain localities in central Brazil are particularly rich in endemic species, probably partly due to speciation as a response to local conditions, and partly to reliction following range contraction of formerly widespread species as a result of climatic changes. The two processes are perhaps also mutually dependent. Haffer (1969) has recently recognised centres of diversity in the Andes and Amazon Basin, for different bird groups. These centres he considers to represent refugia at times when unfavourable climatic conditions caused extensive contraction of the present Amazonian forest. Although little is known of past climatic changes in central Brazil, it seems likely that similar oscillations causing far-reaching vegetational modifications will have also occurred here. Mountain regions, which can never have been completely forest-covered even at times of maximum forest expansion, allow altitudinal migration, and thus provide a buffer against climatic change and, at the same time, due to their physiography, provide a wide range of open habitats suitable to savanna and cerrado species. Climatic changes by causing range expansions will provide opportunities for gene exchange between advancing populations, and by causing range contractions will provide geographical isolation leading to local differentiation by natural selection as well as by genetic drift. It is not surprising therefore that the montane areas of central Brazil support rich biological patterns of variation and speciation, though admittedly the altitudinal range is very slight compared to the Andean chain in the west.

In Minas Gerais, two areas are particularly rich in *Eriope* species; these are Diamantina, which contains four of the eleven species restricted to this and the neighbouring state of Bahia, and the Serra do Cipó with three species, two of which, *E. filifolia* and *E. hypoleuca*, are also known from Diamantina, but nowhere else.

The little-known region in south Bahia, roughly bordered by Jacobina and Morro do Chapeu in the north, and Sincorá and the Rio de Contas in the south, is another area of high endemism. Here *Eriope* seems to have undergone an explosive phase of radiative evolution, probably as a result of unusual environmental factors. Many of the plants in this region are adapted to semi-arid conditions and often show close parallelisms in xeromorphic characters. At least eleven species of *Eriope* are recorded from the area of which six or possibly seven are endemic to it. Many of these have not been refound since their discovery in the early nineteenth century, and it is highly probable, as with *E. exaltata*, that further discoveries will be made.*

Mention should also be made of the monotypic genus *Eriopidion*, with its only species *E. strictum*, formerly included in *Eriope*, but differing in several important respects. This is restricted to an area in the north-east of Brazil in semi-arid habitats, which do not support *Eriope*, *sensu stricto*. The morpho-

---

* The results of the 1974 expedition to Bahia under the joint sponsorship of the Kew Royal Botanic Gardens, and the Academia Brasileira de Ciências have now vindicated this statement. A new species collected at that time, *E. tumidicaulis*, is described in this account, while the rest of the results will appear as a subsequent paper in Kew Bulletin.

logical evidence would suggest that it has recently diverged from the main *Eriope* species cluster.

Adaptive Significance of Floral Characters

One factor which may have played an important role in speciation is the specialisation of the flowers for pollination by solitary bees. The explosive mechanism of these flowers has been described elsewhere (Harley, 1971) and was earlier reported in the closely related genus *Hyptis* by Burkart (1939). The sequence of events is apparently as follows. The flower buds have a characteristically inflated appearance prior to opening. Opening occurs gradually, but rapidly enough to be easily observed, and the corolla then assumes the untriggered position. The concave lower lip is strongly laterally compressed and retains within it the four anthers which are bent down under tension. The lower part of the stamen filaments, the posterior pair of which are densely covered with long hairs, are conspicuous between the lower lip and the throat of the corolla. The lower lip continues to move downwards very slightly, increasing the tension on the stamens still further. The margins of the lower lip are fimbriate, and these processes often appear to interlock, perhaps assisting in keeping the anthers from bursting out prematurely. The flower is now ready to be visited. The bee, perhaps in search of nectar, alights on the lower lip, and immediately the lip flips back with great rapidity. As it does so the stamens, on which the tension has suddenly been released, flick up to deposit a mass of pollen on the underside of the bee's abdomen. This pollen is often in sufficient quantity to be detectable on the bee even in flight. Although disturbed, the bee rapidly returns to visit other flowers. During this procedure, the stigma is apparently unreceptive and remains short and often concealed in the long staminal hairs. After being released, however, the stamens, which initially assume an ascending position, begin to deflex until the stigma is no longer concealed by the hairs. It seems likely that at this stage the stamens act as an alighting platform for visiting bees, thus allowing pollen to be transferred to the stigma (see Frontispiece, Tab. 3751, for details of floral structure).

The uniformity of floral structure in the *Hyptidinae* is perhaps in part due to the limitations set by the mechanism which would be seriously disrupted by any extensive alteration to the pattern. Changes in corolla tube length, however, by controlling the accessibility of nectar, can selectively favour certain insect visitors at the expense of others, without in any way jeopardising the pollination mechanism. It is interesting to compare the very uniform corolla shapes in *Eriope*, in which the tube is broad and almost campanulate except at the base, thus allowing visits by short-tongued bees, with that of the large-flowered *Hyptis* species in which variation in length of the narrow corolla tube has obviously played an important role in pollinator specificity. These *Hyptis* species are visited by a range of longer-tongued bees dependent on the length of the corolla tube, and one group with large red flowers is even adapted to pollination by humming birds. Among species of *Eriope*, the only significant variation in the corolla is one of size, ranging from 4·5 mm. in *E.*

complicata up to 13 mm. in *E. latifolia.** The visiting bees appear to be collecting both pollen and nectar. Corolla size obviously provides a mechanical obstacle to the smaller bees who must trigger the flowers to obtain the pollen, and in this way a degree of pollinator specificity is ensured.

In *Eriope*, the flowers are short-lived, the corolla usually dropping on the same day it opened. Observations on *E. crassipes* (Harley, 1971) indicate that most of the pollination occurs in the morning. Thus diurnal as well as seasonal differences in flowering and pollination could act as a barrier to gene flow. However, information on this point is at present insufficient.

Owing to difficulties in cultivation, only one plant, of *E. hypenioides*, has so far been raised from seed to flowering at Kew. Initial observations show that although self-compatible, this plant does not set seed in the absence of pollinators, unless artificially selfed. In the field, pollination seems to occur almost without fail, so perhaps it is not surprising that there is no secondary means of ensuring seed set in the absence of pollinators. Owing to the extraordinary similarity of floral structure in *Eriope*, it seems probable that the other species behave in the same way,

In *Eriopidion strictum* it may be significant that, apart from a series of characters—persistent bracts, hygroscopic calyx-teeth, triquetrous nutlets and absence of stylopodium, which at once distinguish it from *Eriope*—it also possesses a tubular corolla, presumably adapted to a different class of pollinators from those visiting *Eriope*. Its occurrence in an area at the edge of the range of *Eriope*, perhaps where typical *Eriope* pollinators were scarce, may have made this floral modification necessary for survival. Its selection must have been a very effective barrier to gene flow.

The role of ants as nectar thieves, particularly in the tropics, has been discussed by Faegri and van der Pijl (1971). In *Eriope* and the larger-flowered *Hyptis* species, where pollination depends on a single unrepeatable operation, the problem can be a particularly acute one. As in many other genera, several species possess viscid glandular hairs on the stems of the inflorescence, which presumably act as a deterrent, while the presence of long rigid setae at the base of the stems could also be interpreted as a similar device. A much more striking adaptation to keep away ants I have described elsewhere as the 'greasy pole' syndrome. Here, there is a sudden developmental change which causes the inflorescence stems and the internodes immediately below to become spindly and wand-like, thus easily shaken, as well as being completely glabrous and smooth with a conspicuous waxy coating. The wax is composed of an apparently unique type of granule, only visible under very high magnification, which has the property of preventing ants from gaining a foothold on the stems. This phenomenon is often accompanied by a fistular swelling at the middle of some of the internodes below the inflorescence. At one time these were believed to be some form of insect gall, but this has now been conclusively disproved. They apparently function as a further obstacle to the upward

---

* *E. latifolia* also differs in its pink flowers, which contrast with the violet-blue flowers of all the other species. Occasional pink-flowered individuals of the latter have been found, however. This is apparently due to the loss of an anthocyanin which also affects the colour of the stems and leaves (see under *E. hypenioides*).

progress of the marauding ant. Experiments with the system demonstrate its remarkable success. The 'greasy pole' syndrome occurs in *E. filifolia, E. hypenioides, E. monticola, E. tumidicaulis* and in *E. crassipes* subsp. *trichopoda* and subsp. *cristalinae* but not in subsp. *crassipes*; it is also found in several species of *Hyptis* Sect. *Hypenia*. Its occurrence not only in closely related but also in more distantly related taxa seems particularly significant as it demonstrates how characters of high selective advantage can arise several times from a common ancestral genetic background.

Within the narrower limits set for the genus *Eriope* by Bentham, all species possess a calyx which is turbinate at maturity, and with a throat closed by a conspicuous white tuft of hairs. The calyx is directed downwards by the deflexed fruiting pedicel. These characters would seem to be functionally related, perhaps to ensure that the ripe nutlets do not fall from the calyx until external conditions are suitable. It is significant that none of the three *Eriope* species with an erect fruiting calyx: *E. latifolia, E. hypoleuca* and *E. salviifolia*, possess this conspicuous tuft of hairs. It is either completely absent or only vestigial. Only *E. exaltata* has the calyx directed downwards but has no tuft of hairs in the calyx-throat. Being a relatively tall tree, wind dispersal may be a more important factor than humidity. The only other tree species, *E. latifolia*, possesses winged nutlets very similar to those of *Hyptis* Sect. *Buddleioides*. There seems no room for doubt that this section is very closely related to *Eriope*.

The nutlets of *Eriope* species are usually mucilaginous, in common with most species of *Hyptis*. The mucilage is of the *Salvia* type, see Hedge (1970). That is to say imbibition is followed by rupture of the outer epidermal wall with the extrusion of long gelatinous processes in which are embedded spiral strands of cellulose. Its function may be to anchor the nutlet in the damp soil, grains of which adhere to the viscid mucilage.

In *Eriopidion*, the calyx is more campanulate, with a hygroscopic upper lip to the calyx, closed when dry but opening widely when wet. The tuft of white hairs is replaced by a line of setae which probably also help to obstruct the passage of the narrow triquetrous nutlets when the calyx is closed. The calyx is downward-pointing at maturity. As *Eriopidion* grows in the more arid regions of Brazil where rainfall is less predictable, it is important that seed dispersal should be restricted to periods when water is immediately available for germination.

Field work on biological problems, even in temperate regions, leaves much to be desired. The richness and diversity of the tropical flora presents many fascinating problems, few of which have yet been even perceived. It is to be hoped that in the future this approach will be utilised wherever possible as a vital corollary to more orthodox taxonomic studies.

Generic Limits in the *Hyptidinae*

Field studies on *Eriope latifolia* (Mart. ex Benth.) Harley, formerly included in the genus *Hyptis*, and on *E. exaltata* Harley, sp. nov., which shows characters intermediate between the two genera, at once indicated the need for a reassessment of the criteria used to distinguish them. The initial result of this investigation has been to modify greatly the traditional views on sectional classification in *Hyptis* and its relationship with *Eriope*. In particular, the liquid preservation of flowers in the field has allowed corolla characters to be given due attention for the first time. This has shown that they are not only important, but fundamental to our understanding of relationships within the group as a whole. Corolla characters are not easily interpreted on dried specimens, and the older ones have generally lost their corollas unless they have been glued irrevocably and inextricably on to the sheet.

Some workers have denigrated the value of floral characters (El-Gazzar & Watson, 1967) to the extent of ignoring certain aspects of these. Although often difficult of interpretation on dried material, this seems insufficient grounds for their rejection. In a group with specialised pollination, floral minutiae may have a selective effect on pollinators, which in turn control the patterns of gene exchange. Where these minutiae have been retained almost without variation by a large group of species, one must conclude that they are of considerable adaptive importance. Not only do they indicate relationships between recently evolved species, but they also suggest one of the factors, namely pollinating agents, which may influence evolution by their effect on limiting the size of the breeding unit. The variation in corolla morphology between the larger-flowered sections of *Hyptis* certainly suggests that this is a valid character by which to assess evolutionary relationships.

In Epling's (1949) Revisión del género Hyptis, the three large-flowered sections under consideration here, namely Sect. *Buddleioides* Benth., Sect. *Mixtae* Epl., and Sect. *Hypenia* Mart. ex Benth., are not always very clearly distinguished. Sect. *Buddleioides* is separated on account of its branched hairs. Sect. *Mixtae*, containing only the single species *H. salviifolia* Pohl ex Benth., is distinguished from Sect. *Hypenia* by its *Eriope*-like calyx, with the three partly connate dorsal teeth, and the oblique throat, and also by the campanulate corolla. *Eriope* itself is traditionally characterised by the racemose inflorescence, the turbinate calyx, indistinctly toothed in fruit and with its throat closed by dense white hairs, and by the deflexed fruiting pedicel.

Epling, in the preface to his revision of *Hyptis*, pointed out that the most closely related genera to *Hyptis—Raphiodon, Eriope, Peltodon* and *Marsypianthes—*represent neighbouring groups that are scarcely separated from each other by a larger hiatus than that which separates some of the sections of *Hyptis*. Nevertheless he believed that it would be unfortunate to disturb the *status quo* by making taxonomic changes which would cause unnecessary name changes.

In order to examine relationships between *Eriope* and those species of *Hyptis* with large flowers in lax panicles, a number of characters were selected. The expression of these characters was observed in a range of species representing *Eriope* and *Hyptis* as shown in Table 1. Omitted from this examination was Sect. *Umbellaria*, partly through lack of adequate material, and partly

TABLE 1

Comparison of characters in *Eriope* and large-flowered *Hyptis* species

	Previous taxonomic position	Branched hairs	Raceme (R) or Cyme (C)	Pedicel deflexed in fruit	'Eriope' type calyx	Calyx-throat hairy	Corolla colour pattern (P)	Corolla-tube widened above	Corolla-tube contracted at base	Stylopodium long (L) short (S) or absent	Nutlets winged (W) or triquetrous (T)
Sect. *Hypenia*		1	2	3	4	5	6	7	8	9	10
H. densiflora	Hypenia	−	C*	−	−	−	Cream	−	±	−	−
H. brachystachys	Hypenia	−	C*	−	−	+	Lilac	−	+	−	−
H. macrantha	Hypenia	−	C	−	−	+	Red	−	±	−	−
H. subrosea	Hypenia	−	C	−	−	+	Pinkish	−	±	−	−
H. vitifolia	Hypenia	−	C	−	−	±	Magenta (P)	−	±	−	−
H. salzmannii	Hypenia	−	C	−	−	+	Blue (P)	−	±	−	−
Sect. *Buddleioides*											
H. asperrima	Buddleioides	+	C	−	−	−	Violet (?P)	−	−	S	W
H. arborea	Buddleioides	+	C	−	−	−	Violet (?P)	−	−	S	W
H. cana	Buddleioides	+	C	−	−	−	Violet (?P)	−	−	S†	W
H. conspersa	Buddleioides	+	C	−	−	−	?	−	−	S	‡
H. leucophylla	Buddleioides	+	C	−	−	−	?	−	−	S	?
Eriope											
E. latifolia	Hypenia	+	R*	−	±	±	Pink (P)	+	+	L	W
E. hypoleuca	Hypenia	+	R*	−	±	±	Violet (P)	+	+	L	‡
E. salviifolia	Mixtae	−	R	−	±	±	Violet (P)	+	+	L	−
E. exaltata	—	+	R	+	−	±	Violet (P)	+	+	L	−
E. macrostachya	Eriope	−	R	+	+	+ +	Violet (P)	+	+	L	−
E. hypenioides	Eriope	−	R	+	+	+ +	Violet (P)	+	+	L	−
E. crassipes	Eriope	−	R	+	+	+ +	Violet (P)	+	+	L	−
Eriopidion											
E. strictum	Eriope	−	R	+	±	+	?	−	+	−	T

* Inflorescence congested and somewhat obscure.
† Stylopodium rather long but cylindrical and not tapering as in *Eriope*.
‡ Nutlets flattened and sharply angled, scarcely winged.

because it seems to form, with a few exceptions, a distinct and well-characterised group.

The following ten characters were selected:—

1. Leaf hairs: simple or branched.
2. Inflorescence: racemose or cymose.
3. Fruiting pedicel: erect or deflexed.
4. Calyx shape: *Eriope*-type (see text) or not.
5. Calyx-throat: hairy within or not.
6. Corolla colour.
7. Corolla-tube: widening above or not.
8. Base of corolla-tube: contracted or not.
9. Stylopodium: long and tapering, short and cylindrical, or absent.
10. Nutlet shape.

Character analysis

1. *Leaf hairs*. As stated above, one of the chief diagnostic features of Sect. *Buddleioides* is the presence of branched hairs. However, a microscopical examination of *Hyptis latifolia* Mart. ex Benth. (including *H. heterantha* Benth.) and *Hyptis hypoleuca* Benth., both placed by Epling in Sect. *Hypenia* Subsect. *Densiflorae* Benth., revealed the presence of branched hairs, sometimes in small quantity. Branched hairs also occur profusely on the recently discovered *Eriope exaltata*.

2. *Inflorescence*. The inflorescence of *Labiatae* is typically derived from a series of cymose branches arranged along a racemosely branched axis of indefinite growth. In *Eriope* the cyme has been reduced to a single flower, in which only the paired bracteoles at the base of the calyx indicate its origin from a three-flowered cyme. The result is a simple or little-branched raceme. In other groups the reduction of the cymes has not proceeded so far. In some species of Sect. *Hypenia*, e.g. *H. macrantha* St.-Hil. ex Benth., the flowers are often borne singly, but occasional cymose branches can occur. Where the distinction becomes most difficult is where the internodes of the side-branches become greatly shortened and the inflorescence becomes compact, as in Sect. *Hypenia* Subsect. *Densiflorae*. An analysis of the species of the latter, however, suggests that while *H. densiflora* Pohl ex Benth. and *H. brachystachys* Pohl ex Benth. appear to have cymose sub-units, *H. latifolia* and *H. hypoleuca* have racemose sub-units. This is reflected to some extent in the overall shape of the inflorescence. In this respect it is interesting to compare the inflorescence of *H. leucophylla* Pohl ex Benth. (Sect. *Buddleioides*) with that of *H. hypoleuca*. These two Brazilian species are sufficiently alike for Epling to have confused them on at least one occasion. However, the obpyramidal outline of the clearly cymose inflorescence of *H. leucophylla* contrasts with the pyramidal, racemose inflorescence of *H. hypoleuca*.

In *Eriope exaltata*, the inflorescence is, as in other species of *Eriope*, distinctly racemose.

3. *Fruiting pedicels*. These are deflexed in all the traditionally accepted species of *Eriope*, while, in the recently discovered *E. exaltata*, the pedicels are deflexed to somewhat spreading in fruit. Elsewhere the fruiting pedicels are either erect, as in *H. latifolia*, *H. hypoleuca* and *H. salviifolia*, or spreading.

4. *Calyx shape*. The *Eriope* type of calyx, turbinate in fruit, horny and with indistinct teeth, does not occur in *Eriope exaltata*. However, in the three species *H. latifolia*, *H. hypoleuca* and *H. salviifolia* a clear relationship can be seen. Although the calyx in these is not turbinate, it is somewhat horny, and particularly in *H. salviifolia* the throat is oblique and the three upper teeth are slightly fused at the base. Bentham himself remarks on this similarity. At the same time the close similarity of the fruiting calyx of all four species is undeniable.

The calyx of *Eriope stricta* Benth. differs markedly from those of other members of the genus, notably in the hygroscopic upper lip.

5. *Calyx-throat*. One of the traditional diagnostic features of *Eriope* is the conspicuous dense tuft of hairs in the throat of the fruiting calyx. In Sect. *Buddleioides* the throat is glabrous. The hairiness of the throat is a relative character to some extent. *H. salviifolia*, which is excluded from *Eriope* partly on the evidence of the lack of hair in the calyx-throat, was found on further examination to have a hairy throat, although the hairs are short and sparse, and therefore are very inconspicuous at all times. In this respect it agrees closely with *Hyptis latifolia*, *H. hypoleuca* and *Eriope exaltata*.

The throat of *Eriope stricta* differs from all other species in possessing a line of long coarse hairs.

In the remaining members of Sect. *Hypenia* the situation is less clear. In *Hyptis densiflora* the calyx-throat is glabrous, but elsewhere whitish hairs are conspicuous although these do not conceal the throat as in *Eriope*. Particularly in *Hyptis vitifolia*, it is merely the margin of the calyx which is hairy.

6. *Corolla colour*. Where information is available on this subject it appears that in *Eriope* all species normally possess a violet corolla, characteristically marked with darker lines. Occasional aberrant plants, as in *E. hypenioides* and *E. crassipes*, occur with pink flowers, apparently due to the loss of one pigment, not only in the flowers but in other parts of the plant also. Similarly coloured and patterned flowers occur also in *Hyptis hypoleuca*, *H. salviifolia* and in *Eriope exaltata*. In *Hyptis latifolia* the flowers are apparently always pink and show the typical *Eriope* type of pattern. The flower colour of *Eriope stricta* is unfortunately unknown.

In Sect. *Hypenia*, if *H. latifolia* and *H. hypoleuca* are removed, there is a wide array of flower colour, and the flowers are patterned only in *Hyptis salzmannii* and *H. vitifolia*. Violet flowers are here notable by their absence.

Unfortunately I have not observed living flowers of any member of Sect. *Buddleioides*, and am uncertain whether they are patterned or not. Where flower colour is recorded it is said to be violet.

7. *Corolla tube*. The shape of the corolla is one of the most striking features of the group when seen in spirit material or on the living plant. In *Eriope* species, with the exception of *E. stricta*, the tube is campanulate, widening rapidly above. This is also true of *Hyptis latifolia*, *H. hypoleuca* and *H. salviifolia* as well as *Eriope exaltata*. In *Eriope stricta*, however, the tube is narrowly cylindrical. The corolla tube is also cylindrical in the remaining species of Sect. *Hypenia* and in Sect. *Buddleioides*.

8. *Base of corolla-tube*. In all species of *Eriope*, including *E. exaltata* and *E. stricta* and also *Hyptis latifolia*, *H. hypoleuca* and *H. salviifolia*, the corolla-

tube is abruptly contracted near the base. In Sect. *Buddleioides* this is not so. In the remaining species of Sect. *Hypenia*, however, the situation is less clear. While *H. brachystachys* appears to have a clearly contracted corolla-tube at the base, this is less distinctly so in all the other species.

9. *Stylopodium*. Apart from the subfamilies *Ajugoideae* and *Prostantheroideae*, the *Labiatae* are characterised by the gynobasic style. However, in certain members of the *Hyptidinae* the style is supported by an elongate, more or less conical, structure, here called the stylopodium, which arises between and usually overtops the nutlets. As the stylopodium appears to be derived from receptacle tissue, and its junction with the style is demarcated by a distinct joint, it would appear that the style is technically no longer gynobasic. Epling noted this feature, which he called the columella, but he did not greatly stress its importance, although he noted its constancy in *Eriope* and in a few Sections of *Hyptis*.

The transference of *Eriope trichopes* Epling to *Hyptis* Sect. *Minthidium* (Harley, 1973a) was partly suggested by the lack of the stylopodium in this species.

In the present case, the evidence is most striking, and helps to confirm that *Hyptis salviifolia, H. latifolia* and *H. hypoleuca* should rightly be removed to *Eriope*. In common with *Eriope* they possess a characteristic long, tapering stylopodium. In all other species of Sect. *Hypenia* this is absent. All members of Sect. *Buddleioides* so far examined possess a stylopodium, but, unlike *Eriope*, this is much shorter and more or less cylindrical in shape.

Finally, in *Eriope stricta*, the only original species in the genus which is found to possess many inconsistent characters, the stylopodium is absent. This would seem good evidence for its separation as a distinct, if related, genus *Eriopidion*.

10. *Nutlet shape*. Although most species have fairly uniform rounded or somewhat flattened nutlets, it is notable that the majority of tree species, including most of those members of Sect. *Buddleioides* examined, possess winged nutlets. These are particularly well marked in *Hyptis arborea* Benth. of northern South America, and perhaps the largest Labiate tree, attaining a height of over 20 m. In *Hyptis conspersa* Benth., also in Sect. *Buddleioides*, the nutlets are merely flattened and sharply angled, but this plant is said to grow up to only one to two metres tall.

In the relatively much smaller statured Sect. *Hypenia*, winged nutlets have not been recorded, except in *Hyptis latifolia*, a tree up to c. 4 m. high. The nutlets of *H. hypoleuca* are flattened and perhaps weakly winged, but more surprisingly perhaps, those of *Eriope exaltata* are unwinged, although this is a tree up to 5 m. or more, which grows with *H. latifolia*. The character of winged nutlets seems clearly associated with the tree habit, and perhaps suggests links between *Eriope* and *Hyptis* Sect. *Buddleioides*.

Finally, *Eriope stricta* differs from all other species in its narrowly elongate triquetrous nutlets.

Summary. A review of the above characters strongly supports the removal of *Hyptis latifolia* and *H. hypoleuca* from Sect. *Hypenia*, and of *Hyptis salviifolia*, sole member of Sect. *Mixtae*, to *Eriope*. It confirms the validity of *Hyptis* Sect. *Buddleioides*, but highlights the heterogeneity of the remainder of *Hyptis* Sect. *Hypenia*. Finally, it confirms the decision to separate *Eriope stricta* as a new monotypic genus *Eriopidion*. On the other hand, it is clear that *Eriope* is very closely related to the large-flowered *Hyptis* species. At present it seems better to keep up *Eriope* as a separate genus, perhaps with a view to further dividing *Hyptis* in the future. It is certainly clear that these sections are more closely related to *Eriope* than they are to some of the small-flowered sections of *Hyptis*. This is corroborated by the evidence from chromosome studies which it is hoped to present in a separate paper.

The observations are summarised in Table 1 (p. 14), in which the species are arranged according to their new, modified, classification.

Key to Genera of Hyptidinae in South America

1. Flowers in usually pedunculate cymose heads. Heads arranged variously . 2
   Flowers in lax or congested cymose or racemose panicles .   .   . 5
2. Calyx-teeth subulate, rigid and spinose. The spinose fruiting head falling as a unit at maturity   .   .   .   . *Raphiodon*
   Calyx-teeth rarely rigid and spinose. Fruiting head not falling as a unit at maturity 3
3. Calyx-teeth conspicuously flattened and expanded at tip   . .*Peltodon*
   Calyx-teeth not as above   4
4. Nutlets cymbiform with an involute laciniate margin on inner face
   .   .   .   . *Marsypianthes*
   Nutlets ovoid, flattened or winged, but never as above   .   . *Hyptis*
5. Stylopodium present. Corolla usually violet or pink 6
   Stylopodium absent. Corolla usually of other colours.   .   . 7
6. Flowers in racemes. Corolla-tube constricted below, rapidly widening and becoming campanulate above   .   .   .   .   .   . *Eriope*
   Flowers in panicles in which at least the ultimate branches are cymose. Corolla-tube narrow, never campanulate   .   .   .   . *Hyptis*
7. Nutlets elongate, triquetrous. Calyx in fruit with broad, rounded hygroscopic dorsal lobe. Fruiting pedicels deflexed .   .   . *Eriopidion*
   Nutlets ovoid or somewhat flattened. Calyx in fruit not as above, usually distinctly five-toothed. Fruiting pedicels not deflexed   .   . *Hyptis*

ACKNOWLEDGEMENTS

I wish gratefully to acknowledge the loan of material from the following institutions: Jardin Botanique National de Belgique, Brussels (BR); Chicago Natural History Museum (F); Conservatoire, Geneva (G); Herbarium

19

Bradeanum, Rio de Janeiro (HB); Herbarium of the Komarov Botanical Institute, Leningrad (LE); Botanische Staatssammlung, Munich (M); Missouri Botanical Garden, St. Louis (MO); The New York Botanical Garden (NY); Muséum National d'Histoire Naturelle, Laboratoire de Phanérogamie, Paris (P); Divisão de Botânica do Museu Nacional, Rio de Janeiro (R); Jardim Botânico, Rio de Janeiro (RB); Botanical Department, Naturhistoriska Riksmuseum, Stockholm (S); Instituto de Botânico, São Paulo (SP); Departamento de Botânica, Universidade de Brasília (UB); U.S. National Museum (Department of Botany), Washington (US); Botanisches Institut der Universität, Vienna (W). I should also like to record my thanks to the Keeper of Botany, British Museum (Natural History), London (BM), and to the Curator of the Fielding Druce Herbarium, Oxford University (OXF), for allowing me to make use of their facilities.

Thanks are also due to the Royal Society, Royal Geographical Society and the National Research Council under whose auspices I took part in the Xavantina–Cachimbo Expedition to northern Mato Grosso, Brazil. I should also like to thank Dr. Howard S. Irwin, Administrative Director of the New York Botanical Garden, for his generous invitation to accompany his 1971 expedition to the Brazilian Planalto, and to Professor L. G. Labouriau, formerly of the Departamento de Botânica, Universidade de Brasília, for his kindly assistance, and to Dr. G. Hatschbach, Director, Museu Botânico Municipal, Curitiba, Paraná, Brazil, and the many others especially in Brazil who have provided me with fine modern collections.

Finally, I wish to acknowledge the invaluable assistance from various colleagues at Kew: from Miss C. A. Brighton for providing the cytological information, to Mr. H. K. Airy Shaw for providing the Latin diagnoses, and to Dr. R. K. Brummitt for his advice, particularly on thorny nomenclatural problems. To Miss Mary Grierson I am indebted for the excellent illustrations, and to Mrs. C. M. Barndon for the graphs. To Mr. R. D. Meikle I wish to offer my thanks for his advice in so many ways, including reading through the final manuscript.

REFERENCES

Bentham, G. (1833). Labiatarum Genera et Species: 62–145. Cambridge.
Bentham, G. (1848). In De Candolle, A. P., Prodromus Syst. Nat. 12: 83–143.
Briquet, J. (1895–7). In Engler, A. & Prantl, K., Die natürlichen Pflanzenfamilien 4, 4: 183–380.
Burkart, A. (1937). El mecanismo floral de la labiada *Hyptis mutabilis*. Darwiniana 3: 425–427.
El-Gazzar, A. & Watson, L. (1967). Consequences of an escape from floral minutiae and floristics in certain Labiatae. Taxon 16: 186–189.
Epling, C. (1936). Synopsis of the South American Labiatae. Fedde Rep. Spec. Nov. Beih. 85.
Epling, C. (1949). Revisión del Género *Hyptis* (Labiatae). Revista del Mus. de La Plata 7: 153–497.
Faegri, K. & van der Pijl. L. (1971). The Principles of Pollination Ecology, ed. 2. Oxford.
Haffer, J. (1969). Speciation in Amazonian Forest Birds. Science 165: 131–137.

Harley, R. M. (1971). An explosive pollination mechanism in *Eriope crassipes*, a Brazilian labiate. J. Linn. Soc. (Biol.) 3: 159–164.
Harley, R. M. (1973a). *Eriope horridula* (Labiatae), a member of the Verbenaceae. Notes on New World Labiatae I. Kew Bull. 28: 121–122.
Harley, R. M. (1973b). A Cuban *Hyptis* transferred from *Eriope*. Notes on New World Labiatae II. Kew Bull. 28: 24.
Hedge, I. C. (1970). Observations on the mucilage of *Salvia* fruits. Notes R.B.G. Edinb. 30: 79–95.
van der Pijl (1972). Functional considerations and observations on the flowers of some Labiatae. Blumea 20: 93–103.

PART II

Eriope

Kunth ex Benth., Lab. Gen. et Sp.: 142 (1833); Benth. in DC. Prod. 12: 140 (1848); Epling in Fedde Rep. Spec. Nov. Beih. 85: 188 (1936).

Trees, shrubs, or herbs with woody rootstocks. Stems usually glandular above, or sometimes glabrous with fistulose internodes with glaucous waxy bloom. Leaves opposite, with branched or simple hairs, sometimes glabrous. Inflorescence a terminal often branched raceme. Flowers solitary, remote or congested in the axils of slender caducous bracts. Paired linear bracteoles present near base of calyx. Calyx* shortly campanulate at anthesis, ten-nerved and more or less hairy in throat, five-lobed with broadly deltoid lobes, the three upper often more or less connate. Fruiting calyx tubular, campanulate or usually turbinate, often with obscure teeth, prominently ribbed and somewhat reticulate above; calyx-throat often closed by dense tuft of hairs, occasionally short, sparse and inconspicuous. Fruiting pedicels usually deflexed, sometimes erect. Corolla zygomorphic, five-lobed, violet or rarely pink, with pattern of darker lines on two-lobed upper lip; lower lip cymbiform, compressed, with laciniate or dentate margin, enclosing the anthers under tension and providing an explosive pollination mechanism. Corolla-tube constricted near base, widening rapidly and campanulate above. Posterior (outer) pair of stamen filaments densely clothed with long hairs. Stylopodium at base of style long and tapering, distinctly overtopping the young nutlets. Stigma two-lobed. Nutlets four, winged or more or less flattened.

Key to Species

1. Hairs in calyx-throat absent or inconspicuous. Trees or shrubs .   . 2
   Hairs in calyx-throat forming a conspicuous tuft, at least in fruit. Shrubs
      or herbs 5

2. Leaves with lamina to 35 mm. long, scurfy, with scattered stout branched
      hairs. Flowers in lax ± simple racemes. Pedicels deflexed in fruit
      .   .   .   . 4. *exaltata*
   Leaves with lamina over 35 mm. long, tomentose at least beneath. Flowers
      in ± compound usually congested racemes. Pedicels erect.   .   . 3

3. Leaves narrowly elliptic to elliptic-lanceolate, rarely ovate-lanceolate,
      with acute apex. Petiole usually over 10 mm. long .   . 3. *salviifolia*
   Leaves rotund to broadly elliptic or elliptic-ovate, with rounded or retuse
      apex. Petiole usually less than 10 mm. long 4

---

* In the following account, calyx length measurements are given in a contracted form. Thus, under *Eriope latifolia* (p. 25): Calyx 3·5–5/3–4·5 mm. long, should be interpreted as meaning: Calyx on upper side 3·5–5 mm. long, on lower side 3–4·5 mm. long.

4. Trees to 4 m. with pink flowers. Leaves distinctly tomentose above, grey-tomentose beneath. Nutlets conspicuously winged . . 1. *latifolia*
Shrubs to 3 m. with violet flowers. Leaves with minute appressed hairs, often appearing almost glabrous above, distinctly white-tomentose beneath. Nutlets flattened but not winged . . . 2. *hypoleuca*

5. Flowering stems with at least one internode glabrous above with glaucous bloom (rarely glandular and hairy in the inflorescence), internodes often fistulose 6
Flowering stems hairy throughout and often glandular above, internodes never fistulose nor with glaucous bloom 10

6. Leaves narrowly linear, up to 5 mm. wide, and sometimes curled. Petioles up to 2 mm. long . . . . 11. *filifolia*
Leaves ovate to ovate-lanceolate, 10 mm. wide or more, never curled. Petioles usually over 2 mm. long 7

7. Subshrubs with rugose leaves; lamina with persistent stout-based spreading hairs above, glandular hairs absent . . . 10. *tumidicaulis*
Shrubs or herbs without rugose leaves; lamina eventually subglabrous above, occasionally with small scattered barely perceptible hairs, and with glandular hairs 8

8. Shrubs or subshrubs with leaves grey- or white-tomentose beneath with appressed hairs 9
Herbs with swollen woody rootstock. Leaves glabrous or with scattered hairs beneath 10

9. Stems densely grey-tomentose below with very short spreading hairs, long spreading setae few. Leaf margin long-ciliate. Petioles under 5 mm. long . . . . 13. *monticola*
Stems dull brown below, thinly pubescent with crisped hairs, stalked glands and long spreading setae. Leaf margin not ciliate. Petioles usually over 5 mm. long . . . . 12. *hypenioides*

10. Stems with long spreading setae below. Upper internodes weakly inflated. Leaf bases rounded . . . 16. *crassipes* subsp. *trichopoda*
Stems without long spreading setae. Upper internodes strongly inflated. Leaf bases cuneate or attenuate . . . 16. *crassipes* subsp. *cristalina*

11. Leaves with lamina less than 30 mm. long 19
At least most of the leaves with lamina 30 mm. or more . . . 12

12. Leaves linear or narrowly oblong, up to 5 mm. wide, glabrous . .
. . . . 19. *xavantium*
Leaves not as above, 7 mm. wide or more, with at least some scattered hairs on the veins beneath 13

13. Herbs to 80 cm. tall, often less, with slender stems from swollen woody rootstock. Leaves flat, smooth above, never rugose (though veins on upper surface often prominulous) nor tomentose beneath. Petioles to 8 mm. long with a usually distinct joint at base . . . . .
. . . . 16. *crassipes* subsp. *crassipes*

Shrubs or subshrubs 1–3 m. tall, without swollen rootstock. Leaves often
    rugose and more or less tomentose beneath, if less hairy then either
    leaves scabrid above and more or less conduplicate, or petioles 10 mm.
    long or more 14

14. Stems glandular-hairy, mixed with shorter spreading hairs. Leaves not
    rugose, ± elliptic-ovate, never grey-tomentose, often scabrid above,
    conduplicate about the arching midrib. . . . . 8. *foetida*
    Stems with glandular hairs only in the inflorescence. Leaves usually
    rugose, ± lanceolate or if broader then densely grey-tomentose, scarcely
    scabrid above, flat 15

15. Leaves ovate to elliptic-ovate, less than twice as long as broad, with obtuse
    apex and cordate base . . . . . . 6. *velutina*
    Leaves ovate-lanceolate to narrowly elliptic-lanceolate, more than twice
    as long as broad, with usually acute apex and truncate to attenuate
    base . . . . (*macrostachya*) 16

16. Leaves with lamina usually 100–125 mm. long, 38–52 mm. wide, scarcely
    rugose; lower surface subglabrous. Corolla 10–11 mm. long . .
    . . . . 5. *macrostachya* var. *grandiflora*
    Leaves with lamina usually less than 100 mm. long, usually less than 40
    mm. wide and often rugose; lower surface distinctly hairy. Corolla up
    to 9 mm. long 17

17. Lamina of leaves 32–48 mm. long, less than three times length of petiole
    . . . . 5. *macrostachya* var. *platanthera*
    Lamina of leaves over 45 mm. long, usually more than three times length
    of petiole 18

18. Stems and leaves beneath densely grey- or white-tomentose. Lamina
    narrowly elliptic with usually obtuse to rounded apex . . .
    . . . . 5. *macrostachya* var. *hypoleuca*
    Stems and leaves beneath thinly tomentose to villous. Lamina usually
    lanceolate with acute apex . . 5. *macrostachya* var. *macrostachya*

19. Shrubs to 1·5 m. tall, with slender branching stems. Leaves on the flowering
    shoots with lamina up to 15 mm. long, usually shorter than the inter-
    nodes, with conspicuous glands on upper surface . (*parvifolia*) 20
    Herbs, or if shrubs then leaves crowded on stout stems. Leaves not
    obviously glandular on upper surface or if so then lamina more than
    15 mm. long 21

20. Leaves with short-stalked glands and stout scattered hairs on upper
    surface. Leaf apex usually rounded or minutely cuspidate. Petiole
    (1·5–)3–4 mm. long . . . 9. *parvifolia* subsp. *glandulosa*
    Leaves with sessile glands and minute dense spreading hairs on upper
    surface. Leaf apex acuminate. Petiole 1–3 mm. long . . . .
    . . . . 9. *parvifolia* subsp. *parvifolia*

21. Leaves densely hairy beneath. Petioles up to 6 mm. long . . . 22
    Leaves glabrous beneath or with scattered hairs on the veins and midrib.
    If hairier then herbs with petioles 5–15 mm. long . . . 24

22. Leaves up to 8 mm. wide, coriaceous, subglabrous above. Margins
conspicuously ciliate . . . . . . 15. *polyphylla*
Leaves at least 9 mm. wide, usually much more, densely hairy above.
Margins not ciliate 23

23. Herbs with swollen woody rootstock. Leaves ovate-elliptic to rotund.
Petiole 1·5–4 mm. long . . . . . . 20. *complicata*
Shrubs. Leaves elliptic. Petiole (3–)4–6 mm. long . .7. *alpestris*

24. Shrubs with stout stems 3–5 mm. in diameter with many conspicuous
abscission scars below. Leaves thick and rigid, densely imbricate and
restricted to upper parts of stem . . . . 14. *crassifolia*
Subshrubs or herbs with slender stems scarcely 2 mm. in diameter,
without abscission scars. Leaves not as above, not densely imbricate
nor restricted to upper parts of stem 25

25. Lamina of leaves obovate-lanceolate to elliptic, glabrous. Margin distinctly thickened and cartilaginous, not ciliate . . 17. *obovata*
Lamina of leaves usually lanceolate or ovate-lanceolate, with scattered
hairs at least along the midrib above. Margin not distinctly thickened
and cartilaginous, usually weakly ciliate 26

26. Subshrub with wiry stems usually much branched near base. Long
spreading setae absent. Leaves up to 7 mm. wide, upper surface with
scattered long, broad-based hairs. Nutlets 2–2·5 mm. long 18. *arenaria*
Herb with straight stems usually unbranched below. Long spreading
setae usually present on stems below. Leaves usually at least 10 mm.
wide, upper surface usually becoming subglabrous. Nutlets 2·5–3 mm.
long . . . . 16. *crassipes* subsp. *crassipes* (dwarf variants)

1. **Eriope latifolia** (*Mart. ex Benth.*) *Harley*, comb. nov. Type: Brazil,
Minas Gerais, in deserto ad Serra de San Antonio/Serra Frio, *Martius* s.n.
(M!, holotype).
Hyptis latifolia Mart. ex Benth., Lab. Gen. et Sp.: 135–6 (1833).
Hyptis heterantha Benth. in DC. Prod. 12: 134–5 (1848), *synon. nov.* Type:
Brazil, Bahia, in Serra Jacobina, *Blanchet* 2587 (K!, holotype; K!, BM.!
OXF! isotypes).
Mesosphaerum latifolium (Mart. ex Benth.) Kuntze, Rev. Gen. 2: 526
(1891).
Mesophaerum heteranthum (Benth.) Kuntze, Rev. Gen. 2: 526 (1891).

Small pachycaul tree to c. 4 m. high, with clearly defined trunk with fissured
corky bark and few ascending branches. Young stems grey-brown-tomentose,
long spreading setae absent, but with scattered sessile glands in the upper part,
the leaves crowded towards the apex of the stems. Lamina 40–55(–70) mm.
long, (23–)30–40 (–49) mm. wide, rotund to elliptic-ovate, often gradually
decreasing above (sterile shoots often with much larger leaves), rigid, rugose

and usually concave above, with broadly to narrowly rounded apex and rounded to subcordate base; upper surface thinly to rather densely tomentose with short broad-based hairs and scattered sessile glands; lower surface densely grey-tomentose to almost lanate with a few branched hairs often present; margin crenate with numerous teeth; petiole 3–11 mm. long, or longer on leaves from sterile shoots, indumentum as on the stems. Inflorescence 10–30 cm. long, forming a compound, elongate, terminal panicle; panicle-branches ascending; indumentum as on the upper stems, with scattered sessile glands. Calyx 3·5–5/3–4·5 mm. long, shortly campanulate, tinged dark purple, grey-tomentose especially in lower half, with five subequal broadly triangular acute teeth c. 1–2 mm. long; calyx-becoming 6–9/5·5–8 mm. long in fruit, campanulate, thinly hairy; hairs in calyx-throat short and sparse; pedicel 1–4(–5) mm. long, not elongating, remaining erect in fruit. Corolla c. 5–11(–13) mm. long, pink with white region at base of upper lobes and overlaid with dark red vertical lines. Style pale pink. Anthers c. 1 mm., filaments pale pink. Nutlets 2–3·5 mm. long, distinctly flattened and winged. n = 10.

BRAZIL, Bahia: in campis Serra das Lages, 1818–19, *Martius* s.n. (M!); in campis asperis montanis ad Monte Santo, 1818–19, *Martius* s.n.; Serra Jacobina, *Blanchet* 2587 (BM!, K!, OXF!); Serra da Agua de Rega, c. 26 km. N of Seabra, road to Agua de Rega, cerrado, 23 Feb. 1971, *Irwin, Harley, Smith et al.* 30803 (K!, NY!, UB!) Serra da Agua de Rega, c. 25 km. N of Seabra, on Agua de Rega road, cerrado, 25 Feb. 1971, *Irwin, Harley, Smith et al.* 31085 (K!, NY!, UB!); Serra do Tombador, summit of Morro do Chapeu, c. 7 km. S of town of Morro do Chapeu, on sandstone with sand-filled depressions and crevices, 16 Feb. 1971, *Irwin, Harley, Smith et al.* 32255* (K!, NY!, UB!)
Minas Gerais: Serra de San Antonio/Serra Frio, *Martius* s.n. (M!).
São Paulo: Ypiranga, *Luederwalt* 135. See note below.

Although Epling originally followed Bentham in retaining *Hyptis latifolia* Mart. ex Benth. and *H. heterantha* Benth. as distinct species, there is evidence that he later had doubts. In a subsequent note (Epling in Brittonia 7: 141, 1951) when recording a collection of *H. heterantha* from São Paulo State as listed above, he expressed his reservations as to its distinctness from *H. latifolia*. Unfortunately I have not seen this material, but from an examination of the two Martius collections of *H. latifolia* (*sensu stricto*) from Minas Gerais and Bahia, and of the Blanchet collection of *H. heterantha* from Bahia, together with the recent material collected by myself, I cannot accept that there are two species. The leaf characters used by Epling to distinguish the two are proved valueless now that more material is available. Although the type of *H. latifolia*, from Minas Gerais, has denser panicles of smaller more numerous flowers than in the other gatherings, careful measurements indicate the very wide variation to be found in the material as a whole, often transcending

---

* Collection from which chromosome number obtained.

the diagnostic characters. Further collections from other areas may, however, eventually suggest infraspecific recognition

TAB. 3752. *Eriope latifolia.* FIG. 1, flowering shoot, × ⅔; 2, sterile shoot, × ⅔; 3, flower, × 4; 4, v.s. flower, × 4; 5, flower × 4; 6, calyx, inner surface, × 4; 7, calyx, inner surface, × 4; 8, fruiting calyx, × 4; 9, nutlet, ventral view, × 6; 10, nutlet, dorsal view, × 6; 11, nutlet, side view, × 6; 12, indumentum from inflorescence-axis, × 40. (2 from *Irwin et al.* 31085; 5 & 7 from *Martius* s.n. (type of *E. latifolia*); remainder from *Irwin et al.* 32255.)

2. **Eriope hypoleuca** (*Benth.*) *Harley*, comb. nov. Type: Brazil, Minas Gerais, Diamantina, July 1840, *Gardner* 5102 (K!, holotype); BM!, OXF!, isotypes).
Hyptis hypoglauca Epl. in Fedde Rep. Spec. Nov. Beih. 85: 227 (1936). Type: Brazil, Minas Gerais, prope montem Itambé, *Sello* 1496 (B†, holotype; UCLA!)
Hyptis hypoleuca Benth. in DC. Prod. 12: 135 (1848).
Mesosphaerum hypoleucum (Benth.) Kuntze, Rev. Gen. 2: 526 (1891).

Shrub to 3 m. high, Stems grey with fine velvety tomentose hairs, long spreading setae apparently absent, but with scattered sessile glands throughout. Lamina 35–70 mm. long, 20–50 mm. wide, broadly elliptic to ovate-elliptic, coriaceous with broadly rounded to sometimes weakly retuse apex and rounded to subcordate rarely almost cuneate base; upper surface covered with minute persistent appressed hairs and scattered longer hairs, but older leaves appearing glabrous to the naked eye, midrib and primary veins grey-tomentose above, lower surface densely white-tomentose with very fine, occasionally branched hairs mixed with scattered sessile glands which hardly obscure the incrassate veinlets; margin crenulate with numerous teeth; petiole 6–8 mm. long, with indumentum as on the upper stems. Inflorescence (7–)10–26 cm. long, forming a congested, compound, terminal panicle, narrowly elongate in shape; panicle-branches narrowly ascending; indumentum as on the stems but denser. Calyx 3·5–4 mm. long, shortly campanulate, apparently purple-tinged and densely grey- or white-tomentose to lanate on the tube below, and with five subequal narrowly triangular acute teeth 1–1·5 mm. long; calyx becoming 6–8 mm. long in fruit, tubular, more thinly hairy, with teeth 1·5–3 mm. long (longest in Glaziou material); hairs in calyx-throat sparse or absent; pedicel up to c. 3 mm. long, elongating to c. 4 mm. and remaining erect in fruit. Corolla 7–8 mm. long, violet. Anthers c. 1 mm. Nutlets c. 3 mm., flattened but not winged.

BRAZIL, Minas Gerais: Diamantina, July 1840, *Gardner* 5102 (BM!, K!, OXF!); Rio das Pedras, dans le campo, 1892, *Glaziou* 19705 (K!); Diamantina, Cons. Mata, arbusto 2 m., flor lilaz, dos afluentes rochosos, 12 Aug. 1972, *Hatschbach* 30205 (K!); Jaboticatuba: Serra do Cipó, arbusto 1 m. 70, flor lilaz, dos afloramentis rochosos, 6 Aug. 1972, *Hatschbach* 29998 (K!).

In his Revision, Epling cites *Glaziou* 19705 under both *Hyptis hypoleuca* and *Hyptis leucophylla.* This latter species is vegetatively remarkably similar,

3752

Eriope latifolia (Mart. ex Benth.) Harley

being distinguished most easily by the coarser indumentum of branched hairs on the inflorescence axis. The corolla is also different, having a longer tube not constricted just above the base as it is in *E. hypoleuca*.

TAB. 3753. *Eriope hypoleuca*. FIG. 1, flowering shoot, × ⅔; 2, flower, × 6; 3, calyx, × 6; 4, calyx, inner surface, × 6; 5, calyx, inner surface, × 6; 6, fruiting calyx, × 6; 7, nutlet, ventral view, × 8; 8, nutlet, dorsal view, × 8; 9, nutlet, side view, × 8; 10, indumentum on stem, × 26; 11, portion of leaf, lower surface, × 14; 12, indumentum from outer surface of corolla-lobe, × 40; 13, indumentum from leaf, lower surface, × 40. (5 from *Glaziou* 19705; remainder from *Gardner* 5102.)

(*See page 31*)

3. **Eriope salviifolia** (*Pohl ex Benth.*) *Harley*, comb. nov. Type: Brazil, Minas Gerais, ad Calumbis, *Pohl* 3122 (W, holotype; K!, isotype).
Hyptis salviaefolia Pohl ex Benth., Lab. Gen. et Sp.: 136 (1833).
Hyptis macrotricha Briq. in Ann. Cons. Jard. Bot. Genève 2: 195–6 (1898). Type: Brazil, Minas Gerais, Carandahy, dans le campo, *Glaziou* 13104 (G, holotype; K!, isotype).
Hyptis minensis Glaziou, *nom. nud.*, in Mém. Soc. Bot. France 3: 550 (1911). Cited specimen: Brazil, Minas Gerais, Biribiry, près Diamantina, *Glaziou* 19704 (K!, duplicate).
Mesosphaerum salviaefolium (Pohl ex Benth.) Kuntze, Rev. Gen. 2: 527 (1891).

Shrub to 2 m. high with ascending branches. Stems densely tomentose with short upwardly directed hairs mixed with long spreading setae in the lower part, and with scattered sessile glands above. Lamina (40–)50–65(–70) mm. long, (16–)19–25(–30) mm. wide, narrowly elliptic to elliptic-lanceolate, rarely ovate-lanceolate, rugose, with acute or acuminate rarely somewhat rounded apex, and subcordate to rounded sometimes cuneate base; upper surface sparsely pilose to tomentose with spreading curved hairs; lower surface thinly to densely grey- or white-tomentose; margin serrulate-crenate with numerous teeth; petiole (7–)10–20(–25) mm. long, with indumentum as on the upper stems. Inflorescence (8–)15–30(–35) mm. long, racemose, once-branched* with lower branches occasionally branched again, branches very short, fastigiate, the lowest often hardly longer than the adjacent internode, the whole inflorescence forming a congested spike-like panicle; indumentum as on the stem, with scattered sessile glands. Calyx 2·5–3/2 mm. long, shortly campanulate, tomentose especially in the lower half with grey spreading hairs, becoming 6·5–7/5–6 mm. long in fruit, tubular-campanulate, curved, sparsely hairy with subequal broadly triangular teeth, upper connate at base; hairs in calyx-throat very short and sparse; pedicel 0·5–1 mm., scarcely elongating, remaining erect in fruit. Corolla c. 6–8 mm. long, violet, with whitish area at base of upper lip and pattern of darker lines. Anthers c. 1 mm. Nutlets 2·25–2·5 mm. long, slightly flattened.

---

* The term 'once-branched' here and elsewhere indicates the presence of only one order of lateral branching.

BRAZIL, Bahia: N. of Vitoria da Conquista, 10 July 1964, *Castellanos* 25031 (R!); Rodovia BR-4, na divisa com o Estado de Minas Gerais, 24 Jun. 1965, *Belém* 1161 (UB!)

Minas Gerais: Diamantina district: Sr. de Bonfim, 1816–1821, *St.-Hilaire* 990 (K!); Mendanha, July 1840, *Gardner* 5101 (BM!, K!, OXF!, R!); c. 15 km. NE of Diamantina on road to Mendanha, alt. 1300 m., 26 Jan. 1969, *Irwin et al.* 22583 (K!); Road from Diamantina to Couto de Magalhães, 18 Feb. 1965, *A.P. Duarte* 9084 (HB!, K!); Biribiry, near Diamantina, *Glaziou* 19704 (K!); Diamantina, Chacará das Bicas, alt. 1270 m., 30 Apr. 1931, *Mexia* 5721 (BM!, K!); Diamantina, Agua Limpa, 22 May 1955, *E. Pereira* 1456 (F!, RB!, UCLA!); Diamantina, 20 May 1955, *E. Pereira* 1372 (F!, RB!, UCLA!); Lapinha, c. 18 km. N of Serro on road (MG 2) to Diamantina, alt. 1200 m., 23 Feb. 1968, *Irwin, Maxwell, Wasshausen* 20726 (K!); c. 25 km. SW of Diamantina on road to Gouveia, alt. 1300 m., 16 Jan. 1969, *Irwin et al.* 22098 (K!, UB!); Columbis, *Pohl* s.n. (K!) (the Chapada dos Columbis is c. 75 km. NNE of Diamantina).

S. Minas Gerais: Carandahy, dans le campo, *Glaziou* 13104 (G, K!).

NE Minas Gerais: c. 31 km. NE of Francisco Sá, road to Salinas, alt. 1100 m., 11 Feb. 1969, *Irwin et al.* 23055 (K!); 49 km. from Medina, 9 July 1964, *L. Duarte* 285 (HB!, K!) & *Castellanos* 25009 (R!); side of the Rodovia BR 4, km. 952, Mato Cipó, (possibly in Bahia), 27 June 1968, *Belém* 3733 (K!).

The species is apparently centred on the Diamantina district, with outliers on the state boundary between Bahia and NE Minas Gerais. The single specimen (*Glaziou* 13104) from S. Minas Gerais may be wrongly localised. The Kew sheet of this number merely states 'near Rio de Janeiro.' *E. salviifolia* has been frequently confused with *E. macrostachya*, the two species bearing a remarkable resemblance to each other except in fruit, but distinguished by the presence of glandular hairs in the inflorescence of *E. macrostachya*.

TAB. 3754. *Eriope salviifolia*. FIG. 1, flowering shoot including lower part, × ⅔; 2, calyx, × 6; 3, calyx, inner surface, × 8; 4, corolla, × 6; 5, fruiting calyx, × 6; 6, portion of leaf, lower surface, × 6; 7, portion of lower stem, × 2; 8, portion of inflorescence axis, × 2; 9, nutlet, dorsal view, × 10. (All from *Irwin et al.* 22098.)

(*See page 33*)

4. **Eriope exaltata** *Harley*, sp. nov., affinitate nulla proxima, ab omnibus speciebus habitu arborescente, indumento e trichomatibus abbreviatis ramosis asperatis sistente recedit.

Arbor parva vel frutex, usque 5 m. altus vel ultra. *Caules* juniores pallide hinnulei, trichomatibus abbreviatis ramosis asperatis induti, adspectum furfuraceum praebentes. *Lamina* (18–)25–35 mm. longa, (8–)11–18 mm. lata, elliptica, rigida, scaberula, apice plerumque rotundata et basi cuneata; pagina superior minute furfuracea, pagina inferior (praesertim secus venas) furfuracea atque dimidio inferiore secus costam albido-floccoso-barbata; margo superne crenato-serratus, triente inferiore plus minus integro; petiolus (3–)5–8 mm.

3753

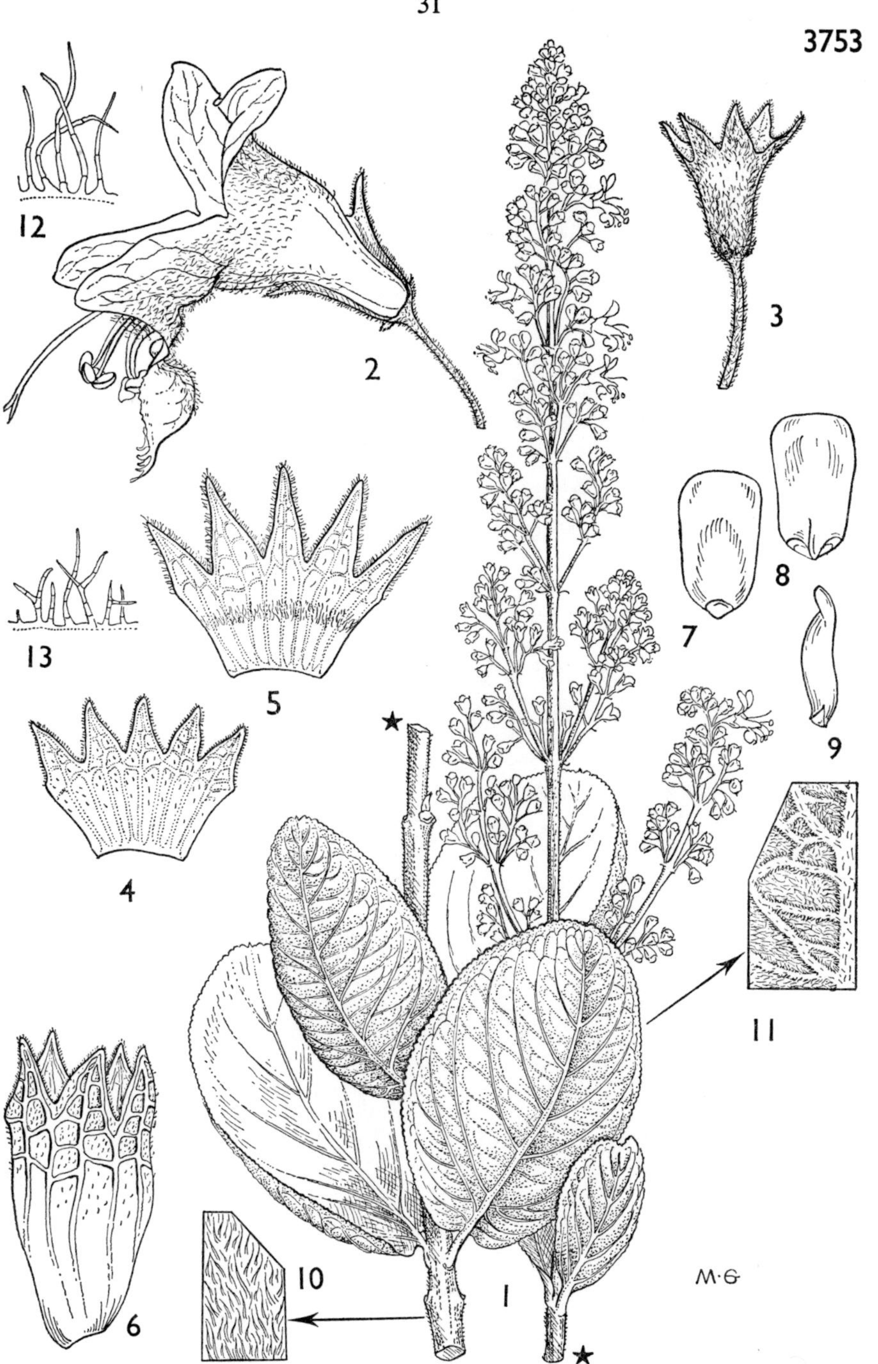

Eriope hypoleuca (Benth.) Harley

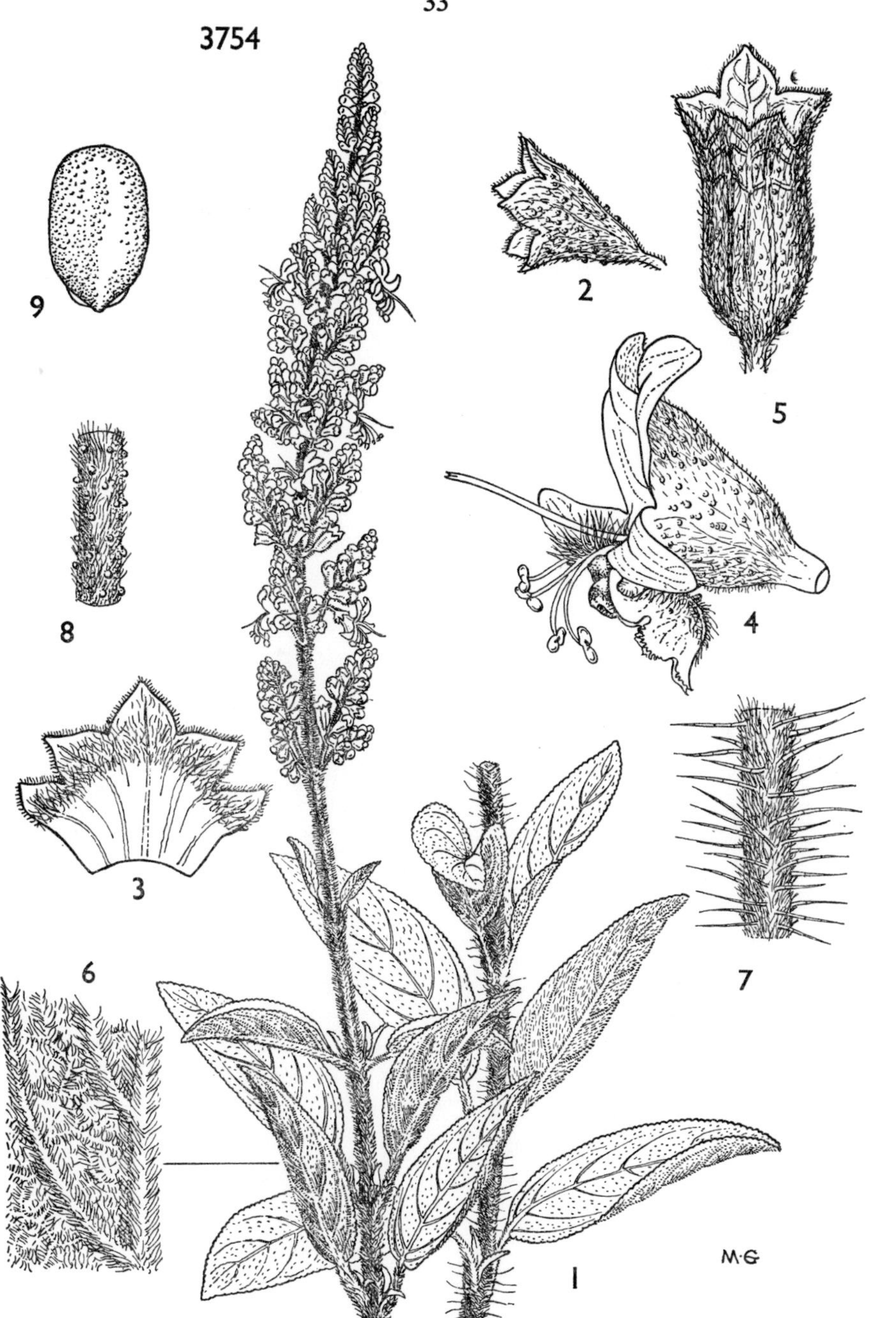

Eriope salviifolia (Pohl ex Benth.) Harley

longus. *Inflorescentia* usque circiter 30 cm. longa, semel ramosa, pilis multi-ramosis fere stellatis cinerascens, pilis glandulosis praeterea numerosis. *Calyx* 3·5–4 mm. longus, primum breviter campanulatus, ± regulariter quinqueden-tatus, demum 6–7·5 mm. longus, campanulato-infundibuliformis dentibus fere aequalibus; pili in calycis fauce sparsi; pedicellus statu fructifero ± de-flexus. *Corolla* 9 mm. longa, pallide violacea. *Nuculi* maturi 2–2·5 mm. longi, complanati.

Small tree or rarely shrub to over 5 m., with clearly defined trunk with longitudinally fissured rather corky bark; stems often sprouting from the base, branches above smooth, grey, very diffuse. Young stems pale fawn, covered with squat, branched, asperate emergences, giving a scurfy appearance, setae absent. Lamina (18–)25–35 mm. long, (8–)11–35 mm. wide, elliptic, rigid, with obtuse usually rounded apex and cuneate base, slightly scabrid, upper surface minutely scurfy, lower surface scurfy especially along the veins, and with white floccose tufts along each side of the midrib in its lower half; margin crenate-serrate in upper two thirds, more or less entire below; petiole (3–)5–8 mm. long. Inflorescence to 30 cm. long, once-branched and forming terminal panicles at the top of the crown of the tree, with greyish indumentum of more slender much branched almost stellate hairs mixed with numerous glandular hairs. Calyx 3·5–4 mm. long, shortly campanulate at first, almost regular, with five subequal broadly triangular acuminate teeth, with much branched hairs mixed with glandular hairs especially on the tube; calyx becoming 6–7·5 mm. long in fruit, narrowly campanulate with more or less regular teeth; hairs in calyx throat sparse; pedicel 1·5–2·5 mm. long, hardly elongating, deflexed or spreading in fruit. Corolla c. 9 mm. long, pale violet-blue, upper lip with broad white region near base of each lobe, with pattern of slightly interrupted dark blue lines. Style very deep violet-blue. Anthers c. 1 mm., dark; filaments and filament hairs white or very pale. Nutlets 2–2·5 mm. long, flattened. n = 10.

BRAZIL, Bahia: Serra da Agua de Rega, c. 26 km. N of Seabra, cerrado, among sandstone rocks, just below summit of small hill, 23 Feb. 1971, *Irwin, Harley, Smith et al.* 30813 (K!, holotype, NY!, UB!, isotypes).

A remarkable and distinct species clearly marked off from most other species of *Eriope* by the indumentum, and by the almost regular calyx, with sparse hairs in its throat, and on a pedicel which is weakly deflexed in fruit. It is perhaps most closely allied to *Eriope salviifolia*.

Its recent discovery on a small hill in Central Bahia, in company with *E. latifolia* and another new Labiate species, *Hyptis irwinii* Harley, indicates the need for further collections in this rich and underworked area.

TAB. 3755. *Eriope exaltata.* FIG. 1, habit, scale indicated; 2, flowering shoot, × ⅔; 3, calyx, side view, × 5; 4, calyx, inner surface, × 5; 5, corolla, × 5; 6, nutlet, ventral view, × 5; 7, nutlet, dorsal view, × 5; 8, portion of leaf, lower surface, × 20; 9, indumentum from stem, × 26; 10, indumentum from outer surface of corolla lobe, × 40. (All from *Irwin et al.* 30813.)

(*See page 37*)

5. **Eriope macrostachya** *Mart. ex Benth.* Lab. Gen. et Sp.: 145 (1833). Types: Brazil, Minas Gerais, in campis ad S. João del Rey et alibi, *Martius* s.n. (M!, lectotype; US!, UCLA!, photo.) Paratypes: Brazil, Minas Gerais, Bois vièrge près le Ponte dos Paulistas, partie orientale de la province, *St.-Hilaire* 1065 (P!, F!); Brazil, Minas Gerais, bords de l'Itacurambi pequeno, *St.-Hilaire* 1680 (P!); 'Brasilia meridionalis', *Sello* s.n. (K!).

a. var. **macrostachya**

E. macrostachya Mart. ex Benth. var. *villosa* Benth., Lab. Gen. et Sp.: 145
  (1833). Type: Brazil, Minas Gerais, in campis elevatis prope Carandai,
  Sello s.n. (B†, holotype; K!, isotype, NY!, US!, UCLA! photo.).
E. macrostachya Mart. ex Benth. var. *grandis* Epling, *ined.*
E. silvatica Tolmatchew in Not. Syst. Herb. Hort. Petrop. 3: 167 (1922),
  synon. nov. Type: Brazil, Minas Gerais, in silvis humidis ripae Rio das
  Velhas, *Riedel & Lund* 2450 (LE!, holotype; NY!, US!, isotypes).

Shrub to 3 m. high, often smaller, with usually numerous ascending branches. Older stems smooth, pale brown in lower part with conspicuous lenticels, younger stems with variable indumentum, usually with short, upwardly directed hairs mixed with longer spreading setae; hairs often tinged reddish-purple in upper part of stem. Lamina (45–)50–80(–115) mm. long, (12–)20–40(–45) mm. wide, ovate-lanceolate to narrowly elliptic-lanceolate, rugose, with acute apex and truncate to weakly attenuate base; upper surface sub-glabrous to sparsely pilose, more densely so along midrib and primary veins, lower surface thinly hairy to tomentose or somewhat villous especially along the veins, glandular on both surfaces; margin weakly serrulate to serrulate–crenate with numerous teeth; petiole (7–)10–20(–40) mm. long, less than $\frac{1}{3}$ length of the lamina. Inflorescence 20–50 cm. long, once-branched with lower branches often branched again, with indumentum as on the upper stems, but mixed with abundant glandular hairs. Calyx 2–3/2 mm. long, turbinate with appressed hairs and stalked glands, becoming 5–8/3–6 mm. long in fruit; hairs in calyx-throat dense often violet- or purple-tinged at first, becoming white; pedicel 1–2 mm., becoming 2–4 mm. long in fruit. Corolla 5–9 mm. long, pale violet to darker blue-violet, with whitish area at base of upper lobes overlaid with pattern of deep blue vertical and transverse lines. Anthers c. 1 mm., yellow. Stigma dark. Nutlets 2–2·5(–2·75) mm. long. n = 10.

BRAZIL, Ceará: Serra do Baturiti, vertente W entre Lagoa e o sertão de Caridade, parte inferior, arbusto 1 m. de altura, flores azuis, 24 Apr. 1909, *Ducke* 2086 (RB!); unlocalised, *Fr. Allemão & de Cysneiro* 1147 (R!).

Minas Gerais: Cerrado, Carapuça, Belo Horizonte, 18 July 1933 *Mello Barreto* 3161 (F!) & 6442 (UCLA!); Belo Horizonte, 9 July 1932, *Brade* 11885 (R!); Slopes of Morro das Pedras, edge of campo, alt. 1035 m., near Belo Horizonte, 12 July 1945, *Williams & Assis* 8011 (BM!, BR!, F!, G!, K!, MO!, NY!, R!, RB!, S!, UCLA!, US!); Estrada de Belo Horizonte a Nova Lima, *Emygdio* 1939 (R!); Serra do Cabral, Joaquim Felicio, Municipio

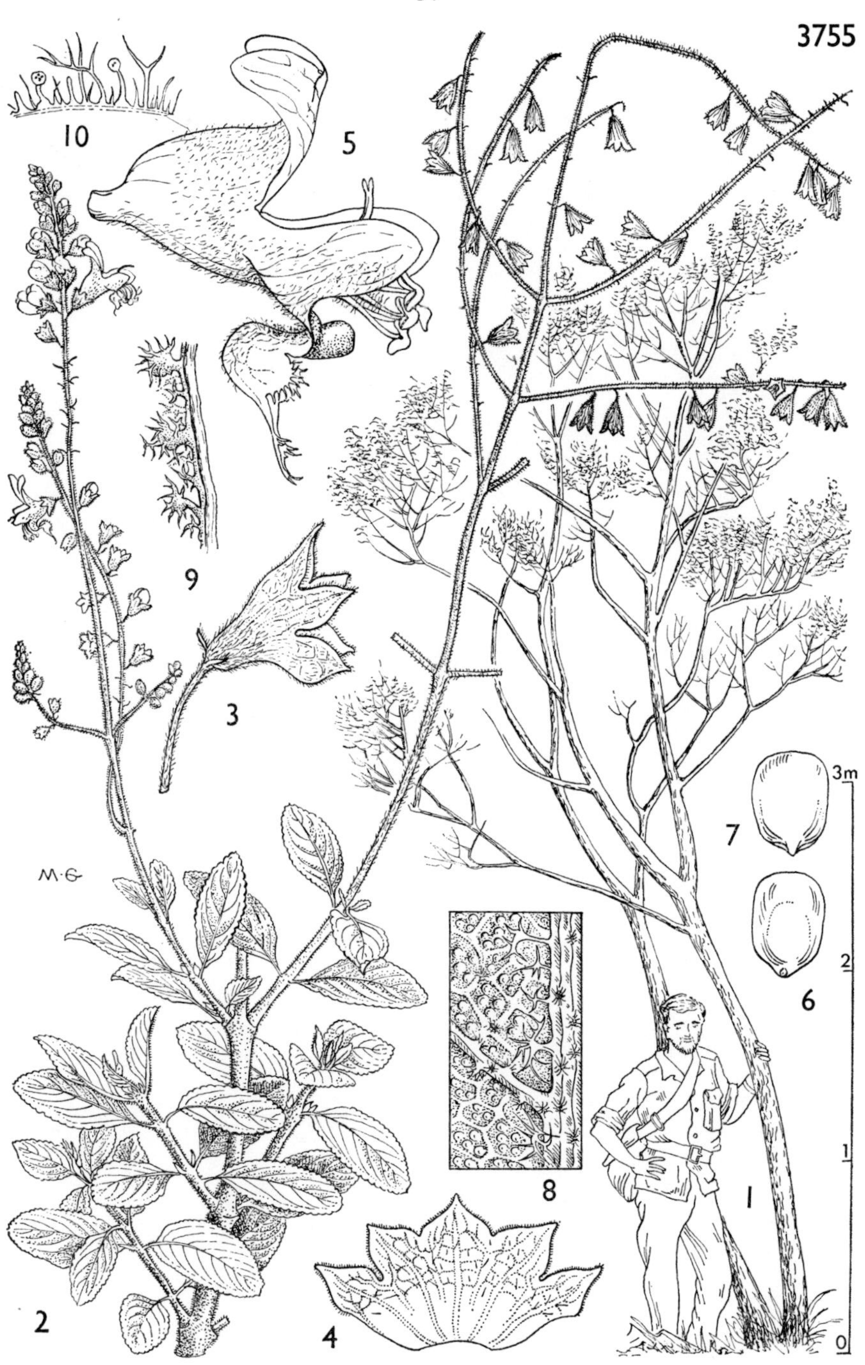

Eriope exaltata Harley

Buenopolis 3 Sept. 1949, *Magalhaes* 4326 (UCLA!); Cachoeira do Campo, Aug. 1839, *Claussen* 177 (G!); Serra do Cipó, Municipio Mato Dentro, 5 Sept. 1952, *Macedo* 3783 (S!, UCLA!, US!); Serra da Piedade, Nov. 1915, *Hoehne* 6585 (R!); south side of Serra da Piedade, iron-rich soil, c. 5 km. N of Caeté, alt. c. 1800 m., 19 Jan. 1971, *Irwin, Harley, Onishi et al.* 28742* (K!, NY!, UB!); Caeté, Sept. 1879, *da Motta* 37954 (R!); entre Barbacena & Barroso, 28 Mar. 1964, *Trinta* 638 & *Fromm* 1714 (K!); Ouro Preto, Sept. 1824, *Riedel* 513 (LE!, NY!, US!); in capoeira, Serra de Antonio Pereira, Ouro Preto, 15 Aug. 1937, *Mello Barreto* 9125 (F!); in silvis humidis ripae Rio das Velhas, Aug. 1834, *Riedel & Lund* 2450 (LE!, NY!, US!); in adscensu montis Serra dos Cabretos, Caldas, 18 June 1854, *Regnell* 1: 317 (BR!, F!, K!, LE!, M!, NY!, S!, UCLA!, US!); Caldas, *Widgren* 1845 (BR!, P!, S!, W!); Lagoa Santa, *Warming* 962 (P!, S!); Serra da Caraça '1816–1821', *St.-Hilaire* 310 (P!); Municipio S. Antonio do Itambe, caminho ao Pico Itambe, 9 Sept. 1971, *Hatschbach* 27514 (K!); bois vièrge, près le Ponte dos Paulistas, '1816–1821', *St.-Hilaire* 1065 (F!, P!); endroits sablonneux de la Serra Negra '1816–1821', *St.-Hilaire* 124 (P!); Caxambú, pr. aeroporto, alt. 900 m., 11 Jun. 1957, *Pabst* 4082 (HB!); Gouveia, Corrego do Tigre, 5 Sept. 1971, *Hatschbach* 27259 (K!); Mindanha, July 1840, *Gardner* 5106 (BM!, K!, W!); Serra da Gamba, 3 May 1888 *Glaziou* s.n. (NY!).

Espirito Santo: Municipio Cach. Itapepemirim, Vargem Alta, Morro do Sal, 4 Aug. 1948, *Brade* 19320 (K!); Vargem Alta, Morro do Sal, 28 Feb. 1965, *Pereira* 9880 (HB!).

Rio de Janeiro: Serra dos Orgãos, Correias, Morro Açú, 29 Oct. 1946, *Pereira & Brade* 16672 (HB!); Petropolis and Itamarity, 20 Sept. 1884, *Glaziou* 15343 (BR!, F!, G!, HB!, K!, LE!, NY!, P!, R!, US!); Theresopolis, *Princess Therese von Bayern* 1926 (M!); Itatiaia, Jardims do Parque Nacional, cultivated, 12 Oct. 1945, *Altimiro & Walter* 44 (RB!).

São Paulo: Mogy das Cruzes, 18 Apr. 1921, *Gehrt* 5482 (SP!, UCLA!); Mogy das Cruzes 1833–1835, *Riedel* 1549 (LE!, NY!, US!); Brigadeiro Tobias, 10 Nov. 1936, *Hoehne & Gehrt* 36721 (SP!); Campos do Jordão, 10 Sept. 1937, *Campos Porto* 3380 (K!).

Paraná: Jaguariahyva, 20 Apr. 1910, *Dusén* 9744 (LE!, NY!, S!); 22 Oct. 1910, *Dusén* 10690 (BM!, S!, US!); 17 Nov. 1914, *Dusén* 15970 (S!); 14 Feb. 1914, *Dusén* 16692 (K!, MO!, S!); Municipio Jaguariaiva, c. 24°08' S, 49°16' W, alt. 810 m., 18 Jan. 1965, *Smith, Klein, Hatschbach* 14760 (US!); Municipio Senges, Fazenda Morungava, Rio do Funil, 14 Dec. 1958, *Hatschbach & Lange* 5367 (UCLA!); 19 Jan. 1965, *Smith, Klein, Hatschbach* 14844 (US!); Municipio Tibagi, Fazenda Monte Alegre, Salto Conceição, alt. 650 m., 3 May 1958, *Hatschbach* 4762 (HB!, UCLA!).

An extremely variable species, even within each of the two common varieties. Plants from Ceará, NE Brazil, may represent another taxon, but more adequate material is needed. Bentham's var. *villosa* appears to grade into var. *macrostachya*, and does not seem worth maintaining.

---

* Collection from which chromosome numbers obtained.

b. var. **hypoleuca** *Benth.* in DC. Prod. 12; 143 (1848). Type: 'In Brasilia', *Sello*
s.n. (K! lectotype; UCLA!, photo; UCLA!, isotype).
E. tomentosa Tolmatchew in Not. Syst. Herb. Hort. Petrop. 3: 167 (1922).
  Type: Brazil, Minas Gerais, São José, in campis siccis petrosis, *Riedel*
  236 (LE!, holotype).

Habit as in var. *macrostachya* but (?) smaller, generally hairier in all its
parts, stems densely villous to lanate in the lower parts; hairs often bluish-
purple tinged. Lamina shorter, elliptic, with usually obtuse to rounded
apex; upper surface thinly tomentose, lower surface densely white-tomentose
to lanate. n = 10.

BRAZIL, Minas Gerais: Municipio of Belo Horizonte, Bento Pires (Eng.
Nogueira), Feb. 1945, *Williams* 5225 (UCLA!); 20 km. S of Belo Horizonte,
vicinity of Lagoa Seca, Feb. 1945, *Williams* 5482 (SP!, UCLA!, US!); Municipio of Nova Lima, 'vargem de Ouro Podre', Serra da Mutuca, alt. 1200–1400
m., 3 June 1945, *Williams & Assis* 7254 (BR!, F!, K!, MO!, NY!, S!, UCLA!);
Serra da Mutuca, alt. 1300 m., *Assis Moreira in Williams* 5742 (UCLA!);
Municipio da Santa Luzia, fazenda da Chicaca, alt. 1100 m., 13 Dec. 1945,
Assis 187 (UCLA!); c. 45 km. SE of Belo Horizonte, Serra do Itabirito,
grazed campo and cerrado, upland valley c. 1500 m. alt., 8 Feb. 1968, *Irwin
et al.* 19555 (K!, UB!); Serra da Moeda, BR-3, 3 Sept. 1964, *Pereira* 9185,
Pabst 8186, *Hatschbach* (HB!); Caraça, caminho da Chacara, 13 June 1884,
Glaziou 15342 (K!, R!); steep, rocky lower slopes of Serra da Caraça, with
plants mostly restricted to soil-filled crevices, alt. c. 1500 m., 23 Jan. 1971,
Irwin, Harley, Onishi et al. 28963,* Serra da Caraça, March, 1892, *Ule* 2668
(R!); Municipio do Ouro Preto, Casa Branca, alt. c. 1200 m., 11 Dec. 1943,
Williams 8115 (UCLA!); in campis Marianae et alibi, April ?, *Martius* s.n.
(M!); Itabiro do Matto Dentro, alt. 3000 ft., 1912, *Hon. C. Baring* s.n. (BM!);
Serra Rola Moça, c. 10 km. E of Barreiro, c. 4000 ft. alt., 15 Jan. 19?, *Irwin*
2456 (F!, NY!, R!, US!); Serra da Rola Moça, 9 Apr. 1951, *Black & Magal-
hães* 51-11968 (UB!); BR-3, Km. 432, 13 Sept. 1964, *Pereira* 9202, *Pabst* 8203
& Hatschbach (HB!); São José, June 1824, *Riedel* 236 (LE!); Carmo do Rio
Claro, fazenda Corrego Bonito, 5 Sept. 1961, *Andrade* 1017 *& Emmerich*
978 (HB!); unlocalised: 1844, *Weddell* s.n. (BR!, P!); 1858, *Weddell* 1395
(G!); ? 1836 *Sello* s.n., ? 1240 (K!, UCLA!).

PARAGUAY: San Luis, between Rio Apa and Rio Aquidaban, 16 Jan. 1909?,
Fiebrig 4861 (BM!, G!, K!); San Salvador, alto Paraguay, 1 Mar. 1917,
Rojas 2806 (SP!); in campis siccis Arrayo Primero, 1907–8, *Hassler* 10854
(G!); prope Concepcion, in campo Ycuapona, Oct. 1901–2, *Hassler* 7679
(BM!, F!, G!, K!, MO!, NY!, P!, S!, W!); unlocalised: *Fiebrig* 5055 (G!).

In its indumentum and leaf-shape this variety seems close to *E. velutina*.
Although not geographically distinct from var. *macrostachya*, it is found

---

* Collection from which chromosome number obtained.

chiefly in the Belo Horizonte–Ouro Preto area of Minas Gerais although outliers, possibly not genetically closely related, occur. In particular, the Paraguayan material is rather different. Further ecological information may reveal differences in habitat preference from the type.

c. var. **grandiflora** (*Epling*) *Harley*, comb. & stat. nov. Type: Brazil, Minas Gerais, ad Serra do Chumbo, *Pohl* 3316 (W!, holotype; UCLA! photo).
E. grandiflora Epling in Fedde Rep. Spec. Nov. Beih. 85: 193 (1936).

Differing from var. *macrostachya* in being generally much less hairy. Leaves larger, with lamina to 125 mm. long, 38–52 mm. wide, thin; upper surface glabrous, lower surface subglabrous, but with sparse crisped hairs more or less restricted to the veins. Petioles 25–48 mm. long. Calyx 3–3·25/2·5 mm. long. Corolla 10–11 mm. long. Anthers c. 1·3 mm.

BRAZIL, Minas Gerais: Serra do Chumbo, *Pohl* 3316 (W!).

Further collections are required to confirm the exact status of this population.

d. var. **platanthera** (*Epling* & *Mathias*) *Harley*, comb. & stat. nov. Type: Venezuela, Tachira, San Cristobal, parte alta, 2020 m., *Garcia-Barriga* 13318 (US!, holotype; UCLA!, photo; NY!, isotype).
E. platanthera Epling & Mathias in Brittonia 8: 313 (1957).

Subshrub to 1 m. high. Older stems smooth, pale brown, with conspicuous lenticels, younger stems with dense slender flexuous hairs; long spreading setae apparently absent. Lamina 32–48 mm. long, 13–26 mm. wide, ovate-lanceolate, often slightly oblong, weakly rugose, with rounded apex and truncate base; upper surface thinly tomentose, lower surface grey-tomentose, inconspicuously glandular; margin serrulate-crenate with numerous teeth; petiole 8–22 mm., usually over ⅓ length of lamina. Inflorescence to 20 cm. long; branches as in var. *macrostachya*, but with shorter glandular hairs. Calyx 2–3/2–2·5 mm. long, turbinate, becoming 6–6·5/4–4·5 mm. long in fruit; pedicel c. 2 mm., increasing to 2·5–4 mm. long in fruit. Corolla 7–9 mm. long. Anthers c. 1 mm. Nutlets 2–2·5 mm. long.

VENEZUELA, Tachira: San Cristobal, alt. 2020 m., 29 Nov. 1948, *Garcia-Barriga* 13318 (NY!, US!).

In most characters this falls within the range of *E. macrostachya*, (perhaps coming closest to var. *hypoleuca*) and therefore is perhaps best treated as subordinate to this species. It differs chiefly in the shorter leaves with a relatively much longer petiole (see Fig. 1) and blunt apex. The indumentum is composed of much finer hairs.

Collection of more individuals may modify these differences, which are based on only a single plant, and therefore it is premature to attempt to evaluate

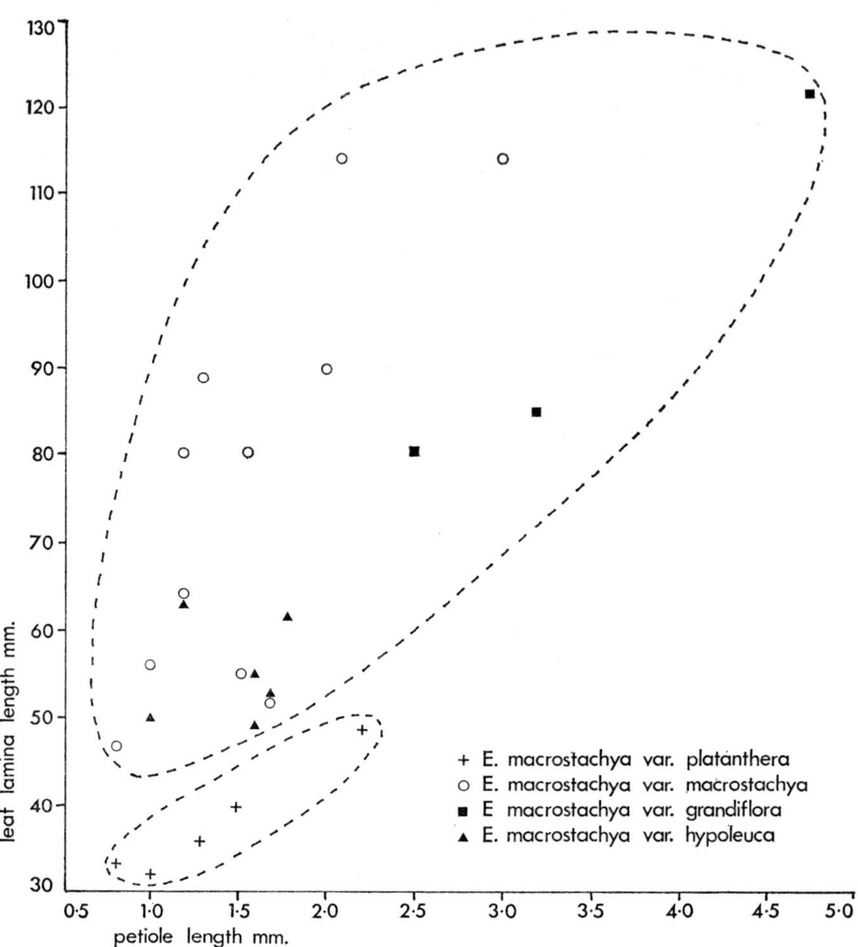

FIG. 1. Leaf lamina/petiole length in *E. macrostachya* (var. *platanthera* is represented by a single gathering).

its exact status. It is interesting to note, however, that this gathering appears to represent a disjunct population separated from the main area of distribution by approximately 2000 miles including the Amazon Basin.

TAB. 3756. *Eriope macrostachya.* FIG. 1, flowering shoot × ⅔; 2, flower, × 3; 3, flower of var. *grandiflora*, × 3; 4, calyx, inner surface, × 6; 5, leaf, upper view, of var. *platanthera*, × ⅔; 6, leaf, upper view, of var. *grandiflora*, × ⅔; 7, leaf, upper view, of var. *hypoleuca*, × ⅔; 8, portion of leaf, upper surface, of var. *hypoleuca*, × 6; 9, portion of leaf; lower surface, of var. *platanthera*, × 6; 10, portion of lower stem, × 2; 11, portion of inflorescence axis, × 2. (All of var. *macrostachya* except where indicated: 1 from *Martius* s.n., lectotype; 2, 4, 10 & 11 from *Irwin et al.* 28742; 3 & 6 from *Pohl* 3316, holotype of var. *grandiflora*; 5 & 9 from *Garcia-Barriga* 13318, holotype of var. *platanthera*; 7 & 8 from *Ule* 2668.)

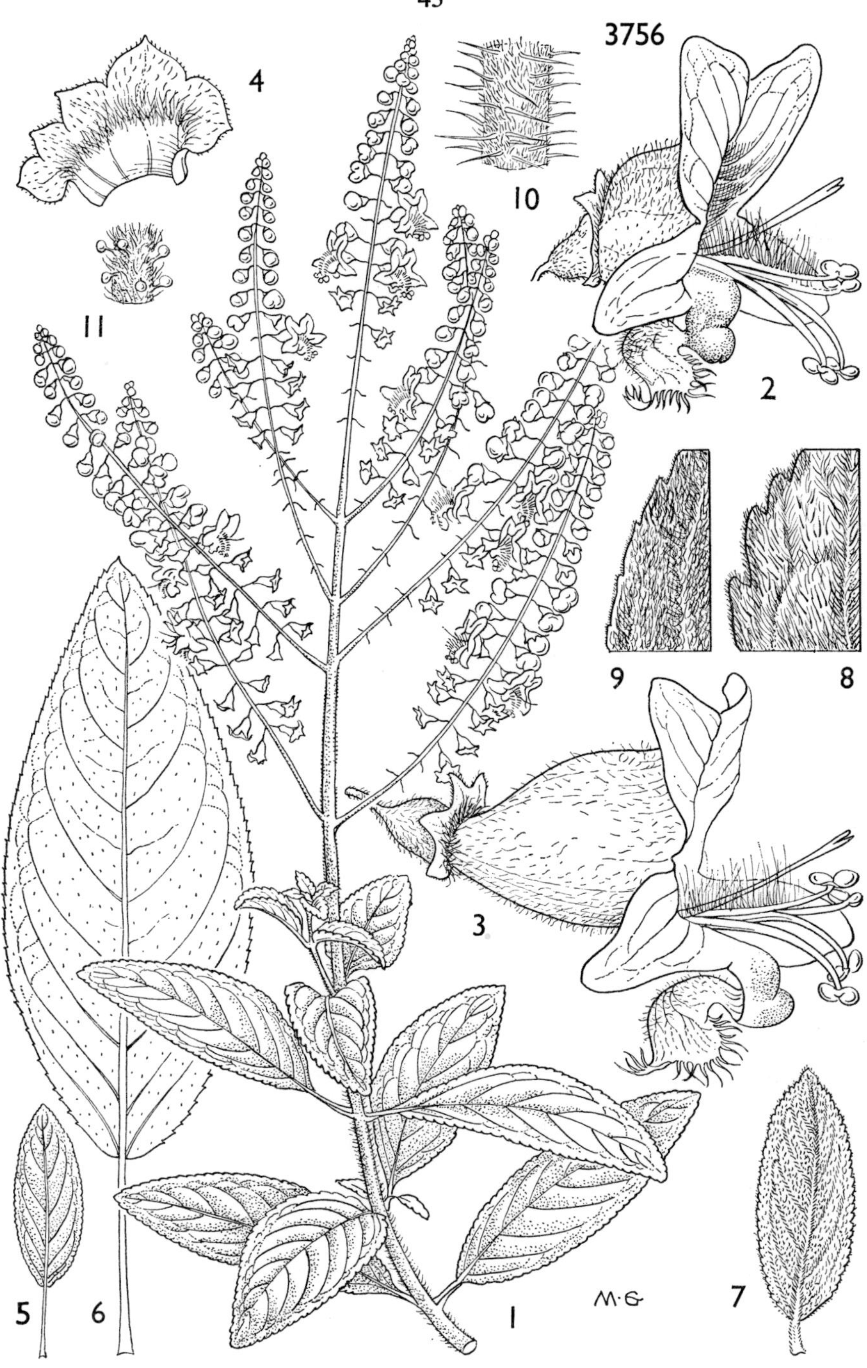

Eriope macrostachya Mart. ex Benth.

6. **Eriope velutina** *Epling* in Fedde Rep. Spec. Nov. Beih. 85: 192 (1936). Type: Brazil, Minas Gerais, Curalinho près Diamantina, *Glaziou* 19684 (B†, holotype; UCLA!; F!, K!, US!, photos; K!, P!, isotypes; UCLA!, photo of Kew sheet).

Shrub, rather sparsely branched, to 1·5 m. Stems densely tomentose to lanate in the lower part with long spreading setae, becoming less densely hairy above with shorter hairs mixed with dense glandular hairs. Lamina (25–)40–70(–80) mm. long, (20–)30–50(–60) mm. wide, ovate to elliptic-ovate,

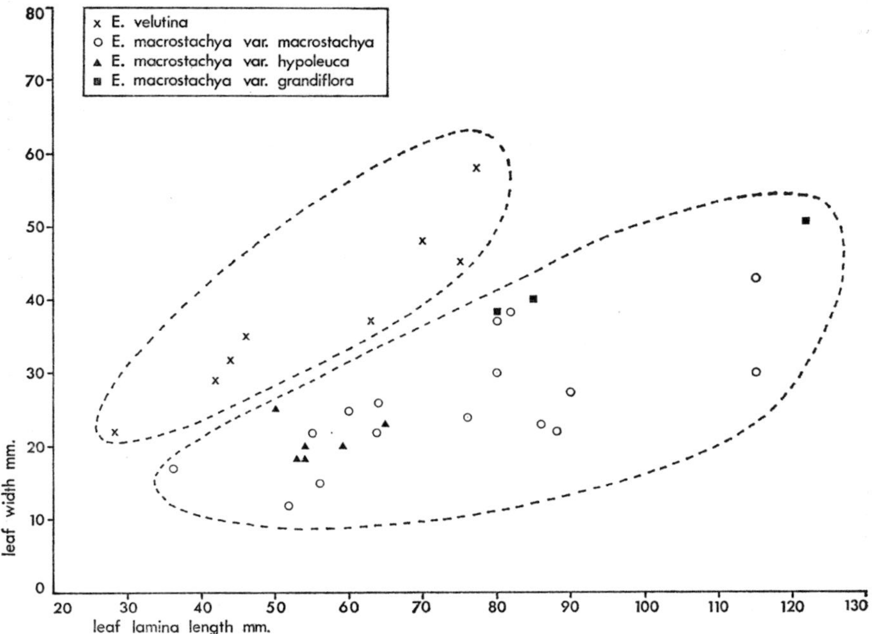

FIG. 2. Lamina length/width in *E. macrostachya* and *E. velutina*.

usually strongly rugose with rounded, obtuse apex and cordate base; upper surface densely grey-tomentose, lower surface densely white-tomentose to lanate between the veinlets (less hairy in specimens from the Serra dos Pirineos), glandular on both surfaces; margin crenate to crenate-serrulate, with numerous teeth; petiole (7–)10–25(–30) mm. long, densely hairy. Inflorescence 20–50 cm. long, once-branched, lower branches usually branched again; indumentum as on the upper stems, with dense glandular hairs. Calyx 2·5–3/2–2·5 mm. long, turbinate with dense whitish appressed hairs and rather few stalked glands, becoming 5–7/3–5 mm. long in fruit; pedicel c. 3 mm. Corolla 5·5–6 mm. long, violet. Anthers c. 1 mm. Nutlets 2–2·5 mm. long.

Brazil, Goiás: Serra dos Pirineos, Dec. 1892, *Ule* 444 (P!, R!, RB!,);*
Serra dos Pirineos, 20 km. N of Corumbá de Goiás on road to Niquelándia,
in valley of Rio Corumbá, c. 1150 m. alt., 18 Jan. 1968, *Irwin et al.* 18757
(K!); Alto da Serra dos Pirineos, 26 Dec. 1969, *Giuletti & Lima* 741 (UB!),
Harley 11498 (K!).

Minas Gerais: Curalinho, près Diamantina, Apr. 1892, *Glaziou* 19684
(K!, P!); Diamantina, 3 June 1955, *Pereira* 1685 (F!, RB!, UCLA!); Diamantina, Agua Limpa, 22 May 1955, *Pereira* 1413 (RB!, UCLA!).

Closely related to *E. macrostachya*, from which it is distinct especially in
leaf shape (see Fig. 2). In the two areas in which *E. velutina* occurs, Diamantina
and the Serra dos Pirineos, *E. macrostachya* is apparently rare or absent.

TAB. 3757. *Eriope velutina*. FIG. 1, flowering shoot, lower part, × ⅔; 2, flowering shoot,
inflorescence, × ⅔; 3, leaf, upper side, from specimen from Goiás, × ⅔; 4, calyx, × 6; 5,
calyx, inner surface, × 6; 6, corolla, × 6; 7, nutlet, dorsal view, × 6; 8, indumentum from
inflorescence axis, × 26. (All from *Pereira* 1413, from Minas Gerias, except for 3 from
Irwin et al. 18757 from Goiás.)

7. **Eriope alpestris** *Mart. ex Benth.*, Lab. Gen. et Sp.: 145 (1833). Type:
Brazil, Minas Gerais, in summo monte Itambé, *Martius* s.n. (M!, holotype).

Shrub of unknown height, with stems densely covered with long spreading
villous to almost setose, broad-based hairs. Lamina (14–)18–23 mm. long,
(9–)11–14 mm. wide, elliptic, rugose, with forwardly directed primary veins,
with obtuse, usually rounded apex and cuneate base; upper surface with
appressed broad-based hairs, primary veins deeply impressed, lower surface
densely villous, primary veins prominent; margin crenate with numerous
teeth; petiole (3–)4–6 mm. long, densely villous to almost setose. Inflorescence,
in the only specimen available, immature, once-branched, with densely
villous to almost setose hairs mixed with numerous glandular hairs. Calyx
immature, densely villous, 2·5–3·5 mm. long approx. Corolla c. 5 mm.

Brazil, Minas Gerais: in summo monte Itambé, *Martius* s.n. (M!).

The type specimen is rather scrappy and immature, as is another gathering:
Lanstyak s.n., Sept. 1945, from Serro do Calixto in Minas Gerais, about 200
km. to the North of Itambé, with which it may be conspecific. Lanstyak's
specimen differs from the type in the much longer setose hairs on the stem,
with numerous shorter villous hairs, and in becoming glandular-hairy above.
The leaves are larger, up to 37 mm. long, 24 mm. wide, with rounded to subcordate base; upper surface with finer hairs, and with numerous glandular
hairs, lower surface densely villous, with finer hairs; margin crenate to
crenate-serrate; petiole up to 10 mm. long.

---

* Sheet in the Herbarium of the Museu Nacional, Rio de Janeiro, (R) with this collection
number and date gives locality as Serra Dourada, presumably in error.

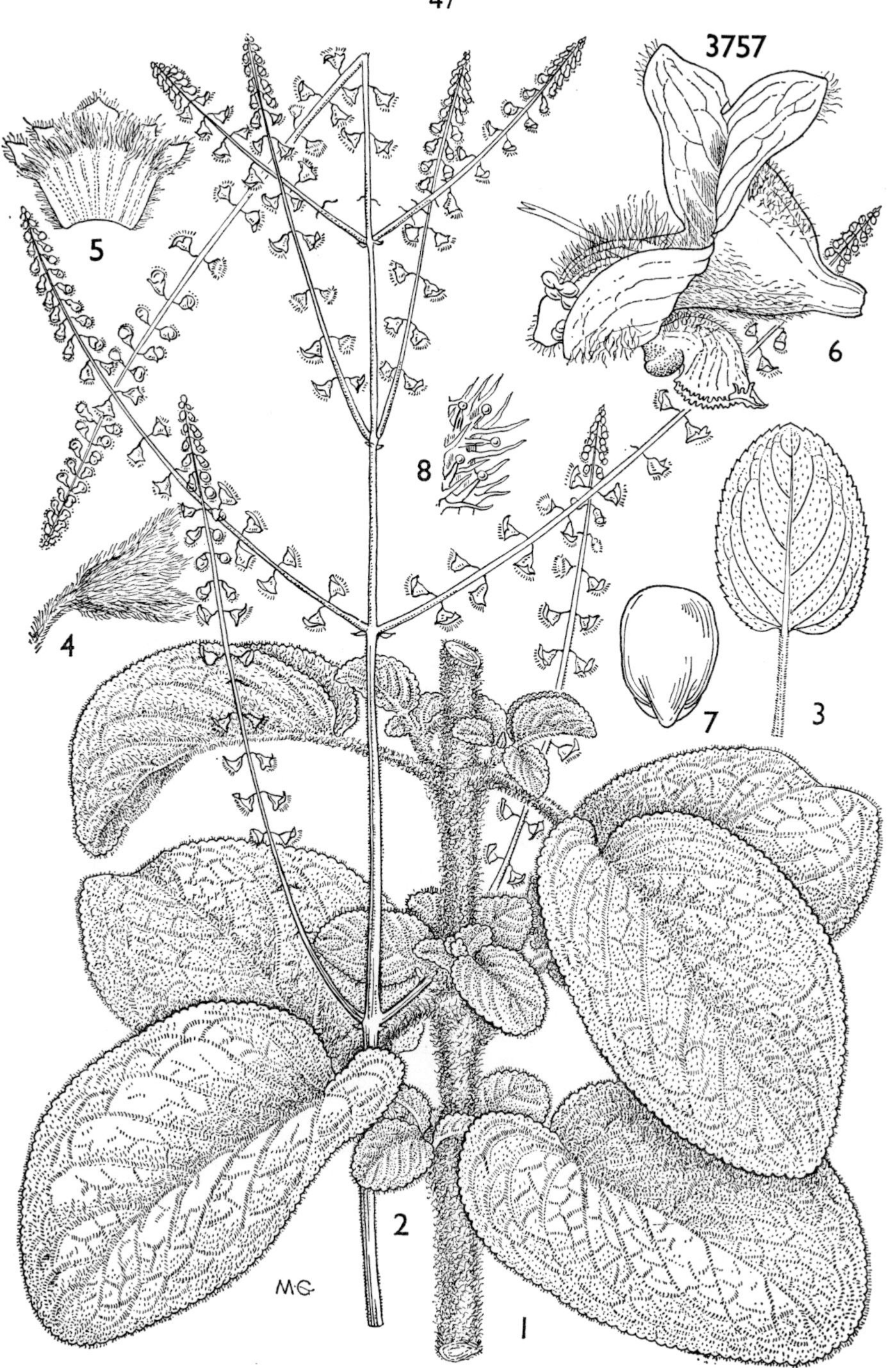

3757

Eriope velutina Epling

Epling wrongly cites as type of *E. alpestris* a Martius specimen 'in campis editis ad Lages, inter Prov. Bahiae, Martius s.n.'. Bentham, however, clearly intended the specimen from Itambé to be the type, and he treats the Bahia plant as var. '*β? glabrior*'. Though not validly published as a variety, this is a very different plant which probably represents a distinct taxon. Further material of all these plants is obviously required.

TAB. 3758. *Eriope alpestris*. FIG. 1, flowering shoot, × ⅔; 2, calyx, × 8; 3, calyx, inner surface × 8; 4, corolla, × 8; 5, leaf, upper view, × 2; 6, portion of leaf, upper surface, × 6; 7, portion of leaf, lower surface, × 6. (All from *Martius* s.n., holotype.)

(*See page 51*)

8. **Eriope foetida** *St.-Hil. ex Benth.*, Lab. Gen. et Sp.: 145 (1833). Type: Brazil, Minas Gerais, entrée du desert occidental de la province, *St.-Hilaire* 1728 (P!, holotype; UCLA!, photo; F!, isotype). Other specimen cited is *E. crassipes*.

Erect foetid shrub to 1 m. with straight spreading stems. Stems throughout with very short dense spreading hairs mixed with glandular hairs, and with long spreading often dense setae in the lower part. Lamina (38–)45–70(–82) mm. long, (20–)25–34(–42) mm. wide, elliptic-ovate, coriaceous, slightly conduplicate about the arching midrib, with broadly acute often cuspidate apex and subcordate base; upper leaf surface ± scabrid, with sparse, short, rigid broad-based hairs, appearing almost glabrous, but with short appressed hairs along the impressed midrib, lower surface densely glandular, pubescent with short erect hairs inconspicuous except along midrib and veins; margin rather irregularly serrate-dentate with numerous teeth; petiole (4–)7–12(–22) mm. long, not jointed. Inflorescence 20–30(–40) cm. long, once-branched, but terminal and axillary inflorescences often forming larger panicles. Indumentum as on stem, densely glandular-hairy. Calyx 2·5–3/2 mm. long, turbinate, densely glandular-hairy and with scattered spreading pilose hairs, becoming 5·5–6·5/4·5–5 mm. long in fruit; pedicel 1·5 mm. becoming 2 mm. long in fruit. Corolla 5–6·5 mm. long. Anthers c. 1 mm. Nutlets c. 2·5 mm. long.

BRAZIL, Goiás: c. 42 km. N of Alto do Paraiso, Chapada dos Veadeiros, 25 Mar. 1971 *Irwin, Harley, Smith et al.* 33153 (K!, NY!, UB!).

Minas Gerais: 'entrée du desert occidental' (apparently to the NW of the Diamantina district, see note below), 1816–1821, *St.-Hilaire* 1728 (F!, P!); between Vieira do Mato and Columbi (c. 100 km. S of Montes Claros), *Pohl* 3170 (BR!, M!, UCLA!, W!).

The distribution of this species has been something of a mystery, but now seems clearer. In Bentham's original description, two specimens are cited, both collected by St.-Hilaire. The first of these is no more localised than: 'entrée du desert occidental de la province ds Minae Gerais'. By reference to

St.-Hilaire's published account of his journeys this seems to be an area to the NW of Diamantina, towards the Goiás border.

The other specimen cited by Bentham is more explicit: 'in pascuis exaratis prope pagum Corumba in regione meridionali provinciae Goyaz'. However, close inspection of the specimen, particularly of the fruiting calyces, and of the stem and leaf indumentum, shows that this is no more than a large individual of *E. crassipes*. Epling selected the first of these two specimens as lectotype, thereby unconsciously ensuring that the generally accepted name is retained for the species now under consideration.

In 1848 in De Candolle's Prodromus, Bentham cites a further specimen, collected by Pohl, purporting to come from Goiás. The sheet from Bentham's herbarium, now at Kew, has the following information: 'inter Vieyra do Mattas et Calumbis prov. Goyaz Pohl'. The only other sheet I have found in which the province is recorded, is one in the Brussels Herbarium in which the following is found: 'inter Vieiras do Mattos et Calumbi Minarum Pohl'. By reference to Martius' account of collectors' itineraries in Flora Brasiliensis, it appears that Pohl travelled between Fazenda Vieira do Mato and sitio Calumbi while in an area c. 100 km. or more south of Montes Claros, to the west of the Rio Jequitinonha, well within Minas Gerais, and not in Goiás.

Finally, the only other sheet of *E. foetida* at Kew, originally in Hooker's herbarium, bears the legend: Pohl 59, Brazil, Herb. Imp. Vienna.

A search among Vienna material of this species, and elsewhere, throws no light on this number, the only Pohl collection of *E. foetida* bearing the number 3170. In view of the frequency with which Pohl specimens are duplicated at Kew, with one sheet from Bentham's and the other from Hooker's herbarium, it seems clear that a wrong number has in this instance been substituted in the latter's specimen, and that this also should be *Pohl* 3170.

Thus we are left with only two original gatherings of *E. foetida*, one by St.-Hilaire and the other by Pohl, both within Minas Gerais, probably within the drainage area of the Rio Jequitaí. This is an area still awaiting good modern collecting.

The recent gathering from Goiás is therefore of particular interest, representing the first true record from the State. It differs in several respects from the Minas Gerais plants, probably chiefly on account of its being new growth after burning. The main differences are its more slender stems, the smaller lamina 25–45 mm. long, 10–16 mm. wide, scarcely coriaceous; densely glandular on both surfaces, upper surface otherwise glabrous, not scabrid, lower surface much less hairy than in the type. Calyx less hairy. Corolla 4·5–5 mm. long. Nutlets c. 2 mm.

Close examination of the upper leaf-surface under high magnification reveals the presence of very sparse broad-based hairs similar to those producing the typical scabrosity, but less well developed, perhaps due to the rapid, lush growth of new leaves on this material. The full field notes from this specimen are as follows:—

Burnt cerrado among rocks. (?) subshrub or herb, previously burnt, remains of old stems up to 1 m. Young stems herbaceous, to c. 75 cm. Stems brittle,

3758

Eriope alpestris Mart. ex Benth.

brown below with long spreading hairs. Leaves dull, rather pale bluish green, paler beneath, folding upwards along the recurving midrib. Foetid, with odour of *Ballota nigra*, and with yellowish glands. Calyces blackish-purple, becoming glossy with dark red-purple lip at maturation. Hairs in calyx-throat white. Corolla rather deep blue-purple with three dark blue lines along each of lobes of upper lip. Anthers pale lilac. Upper filaments deep purple, with long purple hairs, lower filaments white, with very short scattered purple hairs. Style and stigma purple.

TAB. 3759. *Eriope foetida*. FIG. 1, flowering shoot, lower part, × ⅔; 2, flowering shoot, inflorescence, × ⅔; 3, calyx, side view, × 8; 4, calyx, inner surface, × 8; 5, flower, with part of calyx removed to show corolla, × 6; 6, portion of inflorescence-axis, × 4½; 7, portion of leaf, upper surface, × 6; 8, flowering shoot, lower part, × ⅔. (3, 4, 5 & 8 from *Irwin et al.* 33153; remainder from *St.-Hilaire* 1728, holotype.)

(*See page 55*)

9. **Eriope parvifolia** *Mart. ex Benth.*, Lab. Gen. et Sp.: 144 (1833). Type: Brazil, Bahia, in sylvis catingas ad Sincorá, *Martius* s.n. (M!, holotype; UCLA!, photo; M!, ?isotype; UCLA!, US!, photos).

a. subsp. **parvifolia**

E. pallens Benth. in DC. Prod. 12: 143 (1848), *synon. nov.* Type: Brazil, Goiás, dry sandy campos, between Arrayas & São Domingos, *Gardner* 4310 (K!, holotype; UCLA!, photo; K! OXF.!, isotypes; UCLA!, US!, photos).
E. parviflora Steudel, Nom. Bot. ed. II, 1: 587 (1840) err. typogr.

Shrub to 2 m. high, much branched, with dark brown slender stems with conspicuous pale lenticels in the lower part. Stems with usually very short dense spreading hairs; long spreading setae usually present, scattered on the stems but absent in the inflorescence. Lamina 8–12(–15) mm. long, 5–8(–12) mm. wide, ovate, weakly conduplicate about the midrib, with a usually acuminate apex and cordate base; both surfaces minutely puberulent, rarely with longer hairs, densely covered with sessile glands, dull green when fresh, lower surface with short stout hairs on the midrib and veins; margin slightly thickened and obscurely serrate; petiole (1–)1·5–2·5(–3) mm. long. Inflorescence 10–35 mm. long, simply racemose or once-branched, slender; indumentum as on the stems, sparsely glandular. Calyx 2·5–3/1·5–2 mm. long, turbinate, with short rather dense hairs mixed with glandular hairs, becoming (4–)5–6/3–4·5 mm. long in fruit; pedicel c. 2 mm. long, scarcely elongating in fruit. Corolla 5·5–6·5 mm. long, blue-violet, with pale area in throat overlaid by darker pattern of lines. Anthers 1 mm. Nutlets 2–2·5 mm. long.

BRAZIL, Goiás: between Arraias and São Domingos (near the Bahia border), May 1841, *Gardner* 4310 (K!, OXF!).

Bahia: in sylvis catingas ad Sincorá, Nov., *Martius* s.n. (M!); Serra da Batalha (apparently near Santa Rita de Cássia in W. Bahia, on the Piauí

54

border), Sept. 1829, *Gardner* 2929 (K!, BM!); Rio Roda Velha, c. 150 km. SW of Barreiras, alt. 900 m., 15 Apr. 1966, *Irwin, Grear et al.* 14917 (K!, NY!); Rio Piau, c. 150 km. SW of Barreiras on road to Posse, alt. 850 m., 12 Apr. 1966, *Irwin, Grear et al.* 14676 (K!, NY!); 13 Apr. 1966, *Irwin, Grear et al.* 14738a (K!); Cerrado just N of Roda Velha, 13 Feb. 1971, *Irwin, Harley, Smith et al.* 30237 (K!).

Most of the material of this subspecies shows little variation, although the presumed isotype sheet in Munich is particularly hairy. For other more glandular material see remarks under subsp. *glandulosa.*

TAB. 3760. *Eriope parvifolia* subsp. *parvifolia.* FIG. 1, flowering shoot, × ⅔; 2, leaf, lower view, × 3; 3, leaf, lower view, × 3; 4, calyx, side view, × 6; 5, calyx, inner surface, × 6; 6, corolla, side view, × 6; 7, fruiting calyx, side view, × 6; 8, portion of lower stem, × 2. (All from *Irwin et al.* 30237, except for 3, *Gardner* 4310, and 8, *Irwin et al.* 14917.)

(See page 57)

b. subsp. **glandulosa** *Harley,* subsp. nov., subsp. *parvifoliae* proxima, sed habitu magis erecto, indumento e glandulis viscidis stipitatis sistente, foliis late elliptico-ovatis usque rotundatis apice rotundato et basi late cuneato diversa.

Aromatic and somewhat viscid shrub often with short spreading leafy shoots from near the base, and flowering stems to 1(–1·5) m. high, straight, erect to suberect, little branched except for a few short ascending branches below. Stems with very short dense spreading hairs and short-stalked glands becoming abundant in the upper part; setae absent. Lamina (5–)9–13 mm. long, (4–)7–10 mm. wide, broadly elliptic-ovate to rotund, very weakly conduplicate, rather fleshy, with rounded, often minutely cuspidate apex and broadly cuneate base; both surfaces pale green when fresh, densely glandular-viscid with short-stalked glands, and with scattered short, stout hairs; margin crenate; petiole (1·5–)3–4 mm. long. Inflorescence to 40 cm. long, once-branched, indumentum as on the stems but with dense short-stalked glands. Calyx 2·5/2 mm. long, turbinate, glandular-hairy and with scattered short hairs, becoming 5–6/3·5–4·5 mm. long in fruit. Corolla 5·5–7 mm. long, pinkish-violet with weakly striate markings on the upper lobes, throat paler overlaid by darker pattern of lines. Anthers 1 mm. Nutlets 2–2·5 mm. long. n = 10, 2n = 20.

BRAZIL, Bahia: Serra do Tombador, summit of Morro do Chapeu, c. 7 km. S of town of Morro do Chapeu, on sandstone, in sandfilled depressions and crevices, 16 Feb. 1971, *Irwin, Harley, Smith et al.* 32258* (K!, holotype; NY!, UB!, isotypes); Municipality of Morro do Chapeu, sandy soil, sedimentary-metamorphic basement, over 1000 m. alt., no trees, Velloziaceae, Lythraceae, Melastomaceae etc., 20 Apr. 1944, *Schery* 600 (UCLA!).

---

* Collections from which chromosome numbers obtained.

3759

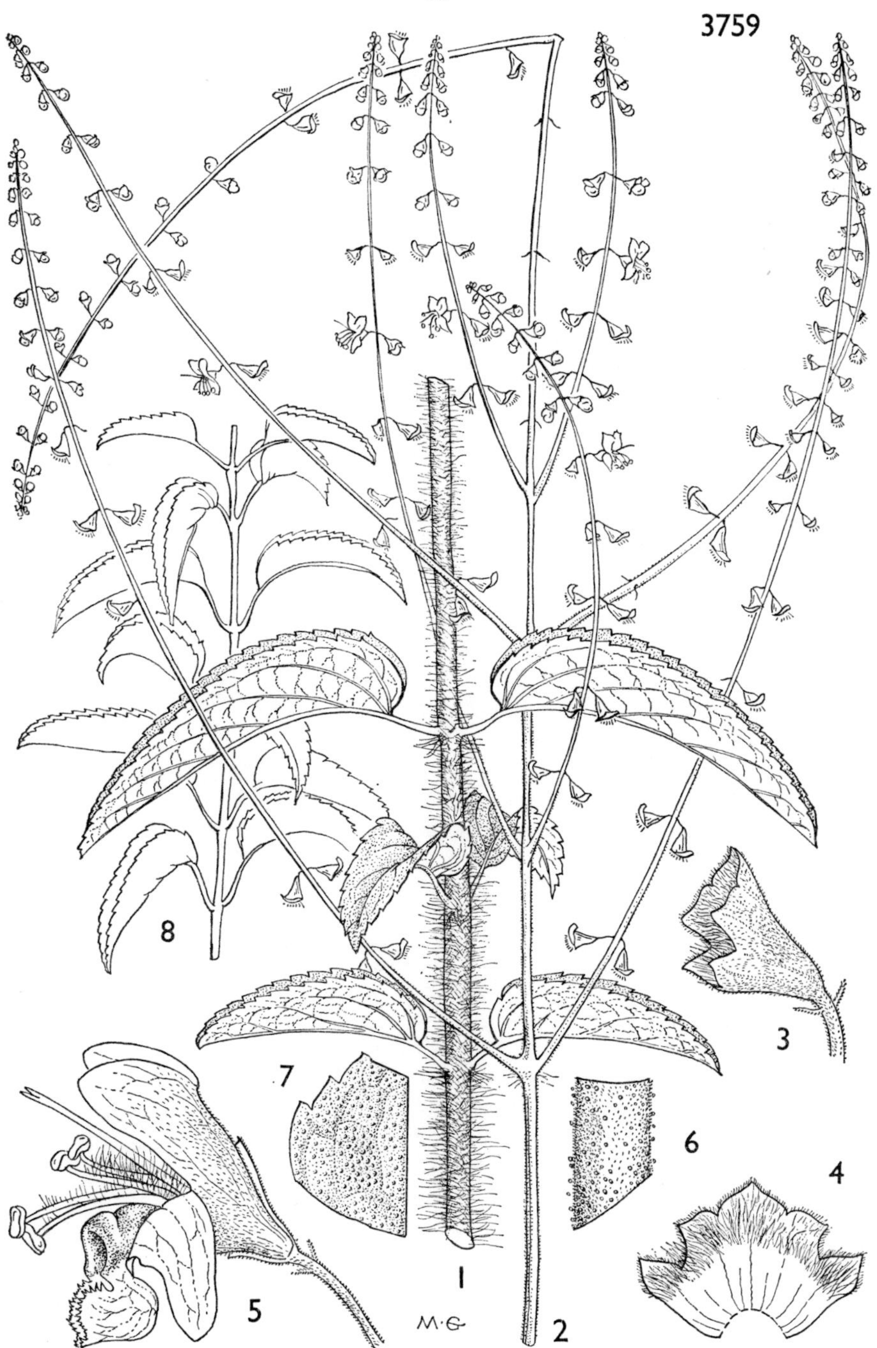

Eriope foetida St.-Hil. ex Benth.

3760

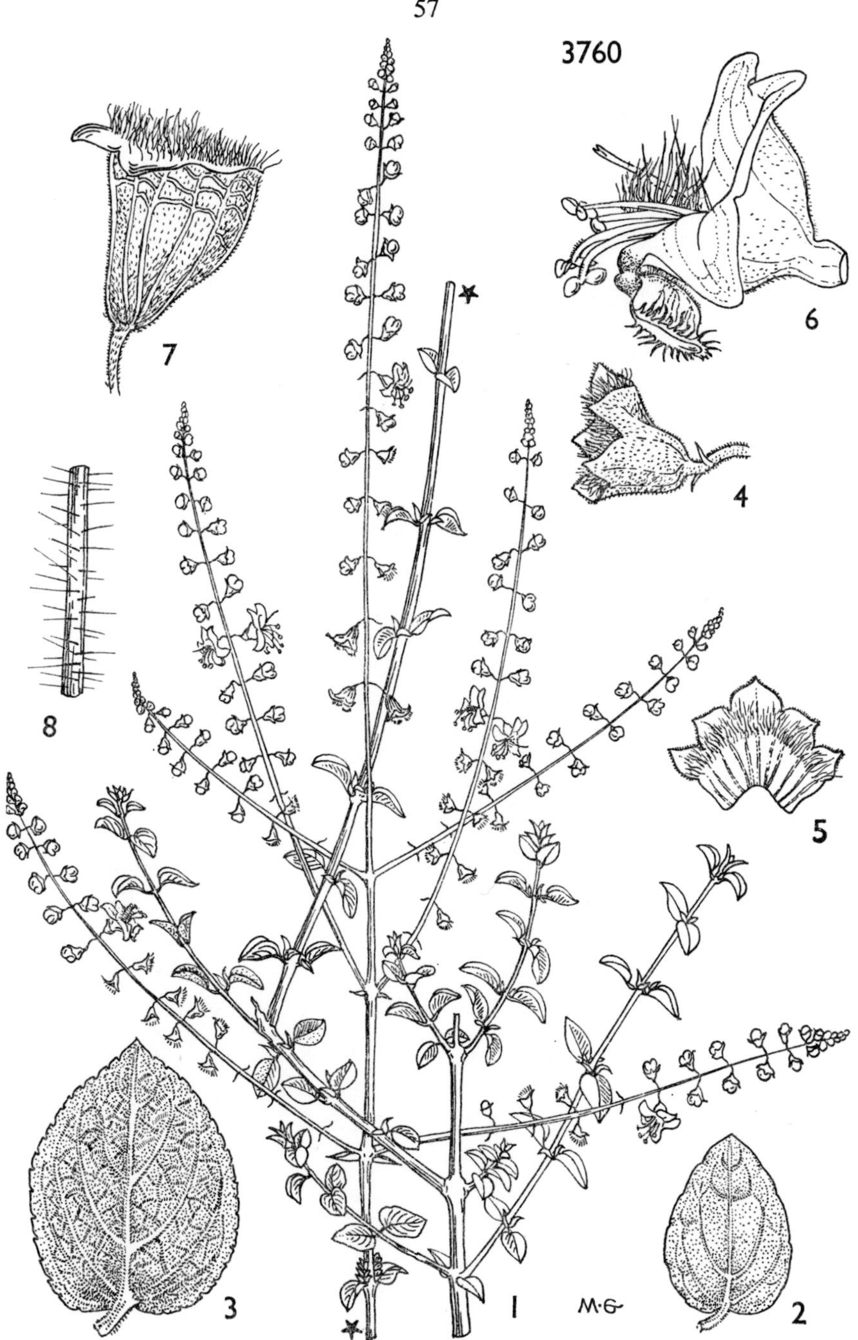

Eriope parvifolia Mart. ex Benth. subsp. *parvifolia*

59

This species occupies a range of habitats in the vicinity of Morro do Chapeu, from often exposed areas at higher altitudes, where it is generally of smaller stature, to sheltered sites, sometimes under small trees, as by the Rio do Ferro Doido, c. 18 km. E of the town of Morro do Chapeu.

Irwin et al. 23349 from Minas Gerais: Grão Mogol, alt. 950 m., 16 Feb. 1969, apparently belongs here. The Kew sheet suggests the same habit as the plants from Bahia, but it differs in its stems with long setae in the lower parts, the subsessile leaves, lamina to 6 mm. long, often broader than long, with rotund apex and cordate base, with numerous stalked glands on the upper surface only.

Three other gatherings from Minas Gerais, Diamantina district, approach ssp. *parvifolia* in leaf shape but show the characteristic indumentum of ssp. *glandulosa* except in the presence of the spreading setae on the lower parts of the stems. These are *Irwin et al.* 22529 (K!), vicinity of Datas 24 Jan. 1969; *Brade* 13646 (RB!), Diamantina 1400 m. June 1934; *Hatschbach* 27298 (K!), Serra do Espinhaco, municipio de Gouveia 6 Sept. 1971. Further material from Minas Gerais is obviously needed before the status of these can be investigated properly.

TAB. 3761. *Eriope parvifolia* subsp. *glandulosa*. FIG. 1, habit, $\times \frac{1}{10}$ approx.; 2, flowering shoot, inflorescence, $\times \frac{2}{3}$; 3, flowering shoot, lower part, $\times \frac{2}{3}$; 4, portion of stem, lower part, $\times 3$; 5, portion of inflorescence-axis, $\times 3$. (All from *Irwin et al.* 32258, holotype of *E. parvifolia* Mart. ex Benth. subsp. *glandulosa* Harley.)

(See page 61)

TAB. 3762. *Eriope parvifolia* subsp. *glandulosa* (*aff.*). FIG. 1, flowering stem, $\times \frac{2}{3}$; 2, calyx, side view, $\times 8$; 3, calyx, inner surface, $\times 8$; 4, corolla, $\times 8$; 5, fruiting calyx, $\times 6$; 6, leaf, upper view, $\times 3$; 7, leaf, lower view, $\times 3$; 8, leaf, upper view, $\times 3$; 9, leaf, upper view, $\times 3$; 10, portion of leaf, upper surface, $\times 16$; 11, portion of stem, lower part, $\times 3$. (1–4, 6, 7 & 11 from Minas Gerais, *Irwin et al.* 23349, a plant showing characters intermediate between subsp. *parvifolia* and subsp. *glandulosa*, 5, 8 & 10 from *Irwin et al.* 32258, holotype of *E. parvifolia* Mart. ex Benth. subsp. *glandulosa* Harley, 9 from Minas Gerais, *Irwin et al.* 22429, showing characters intermediate between subsp. *parvifolia* & subsp. *glandulosa*.)

(See page 63)

10. **Eriope tumidicaulis** *Harley* sp. nov., ab aliis speciebus characteribus sequentibus differt. Frutex vel suffrutex foliis rugosis ovatis vel ovato-lanceolatis; folia pagina supra pilis brevibus patentibus basi incrassatis vestita, subtus sparse vel dense tomentosa pilis patentibus curvatis obtecta. Inflorescentiae internodium infimum conspicue gibbosum glabrum pruinosum.

Frutex vel suffrutex 3 m. attingens sed plerumque humilior, exilis, leniter aromaticus. Caules inferne sordide brunnei, lenticellis conspicuis ornati, dense pubescentes pilis crispis patentibus setisque longis patentibus dispersis vestiti. Lamina (17–)20–35(–45) mm. longa, (10–)12–18(–26) mm. lata, ovata vel ovato-lanceolata, rugosa, apice acuta vel nonnumquam obtusa, basi leviter cordata; supra pilis brevibus patentibus basi incrassatis vestita, subtus sparse vel dense tomentosa pilis patentibus curvatis obtecta; costa et venae primariae subtus plerumque pilis longioribus ornatae, utrinque glandulis sessilibus haud conspicuis obtectae, marginibus serrulato-crenatis leviter undulatis arte dentatis. Petiolus 3–10(–18) mm. longus, subtus ad instar costae pilosus. Inflorescentia 20–50 cm. longa, simpliciter (haud repetite) ramosa, internodio infimo (vel eiusdem duobus inferioribus) glabro, purpureo tincto pruinoso gibba excavata spherica proviso. Calyx 2–2·5/1·5–1·75 mm. longus, sub fructu 5·5–6/4 mm. longus, turbinatus, atropurpureus, pilis plus-minusve adpressis et glandulis sessilibus vestitus, ad faucem pilis albis ornatis. Pedicellus 1 mm. longus, sub fructu 2 mm. attingens et deflexus. Corolla 4–5 mm. longa, saturate rubro-purpurea, lobis superioribus ad basin pallidioribus eleganter fuscostriatis. Antherae c. 1 mm. longae. Nuculae c. 2–2·5 mm. longae.

Weakly aromatic, spindly shrub or subshrub up to 3 m. but usually much smaller. Stems below dull brown with conspicuous elongate lenticels, densely pubescent with short crisped spreading hairs mixed with scattered long spreading setae. Lamina (17–)20–35(–45) mm. long, (10–)12–18(–26) mm. wide, ovate to ovate-lanceolate, rugose, with acute or sometimes obtuse apex and weakly cordate base, upper surface with short, spreading, stout-based hairs, lower surface thinly to densely tomentose with spreading, curved hairs; hairs on the midrib and primary veins beneath usually distinctly longer; both surfaces with scattered rather inconspicuous sessile glands; margin serrulate-crenate, slightly undulate with many teeth; petiole 3–10(–18) mm. long, hairy as on the midrib beneath. Inflorescence 20–50 cm. long, once-branched, with lowest internode, or sometimes with lowest two, glabrous, purple-tinged with white waxy bloom and a more or less spherical swelling, hollow within. Upper internodes of inflorescence with spreading hairs and scattered sessile glands. Calyx 2–2·5/1·5–1·75 mm long, turbinate with more or less appressed hairs and sessile glands, blackish purple, hairs in throat white, becoming 5·5–6/4 mm. in fruit. Pedicel 1 mm. becoming 2 mm. long and deflexed in fruit. Corolla 4–5 mm. long, deep reddish-purple, paler at base of upper lobes with pattern of darker lines. Anthers c. 1 mm. Nutlets c. 2–2·25 mm. long. n = 10.

BRAZIL. Bahia: Serra do Curral Feio, 8 km NW of Lagoinha (5.5 km. SW of Delfino) on the road to Minas do Mimoso, 41°17′ W, 10°24′ S, caatinga/cerrado, frequently burnt and cut-over, alt. c. 850 m., spindly shrub to 3 m., often smaller, stems brown below with conspicuous scattered lenticels, leaves weakly aromatic, rather dull green above, paler beneath, rugose, upper stems on flowering shoots glaucous with waxy bloom, and usually with only a singly conspicuously swollen internode, calyx blackish purple, hairs in throat white, corolla deep, slightly reddish-purple, paler in base of upper lobes with darker pattern of lines, fruiting calyx deflexed, 5 Mar. 1974. *R. M. Harley et al.* 16770 (RB! holotype; CEPLAC*!, K!, MO!, NY!, P!, U!, UCLA!, US!, isotypes). Serra do Curral Feio, 16 km. NW

---

* Herbarium of the Centro do Pesquisas do Cacau, Itabuna, Bahia.

3761

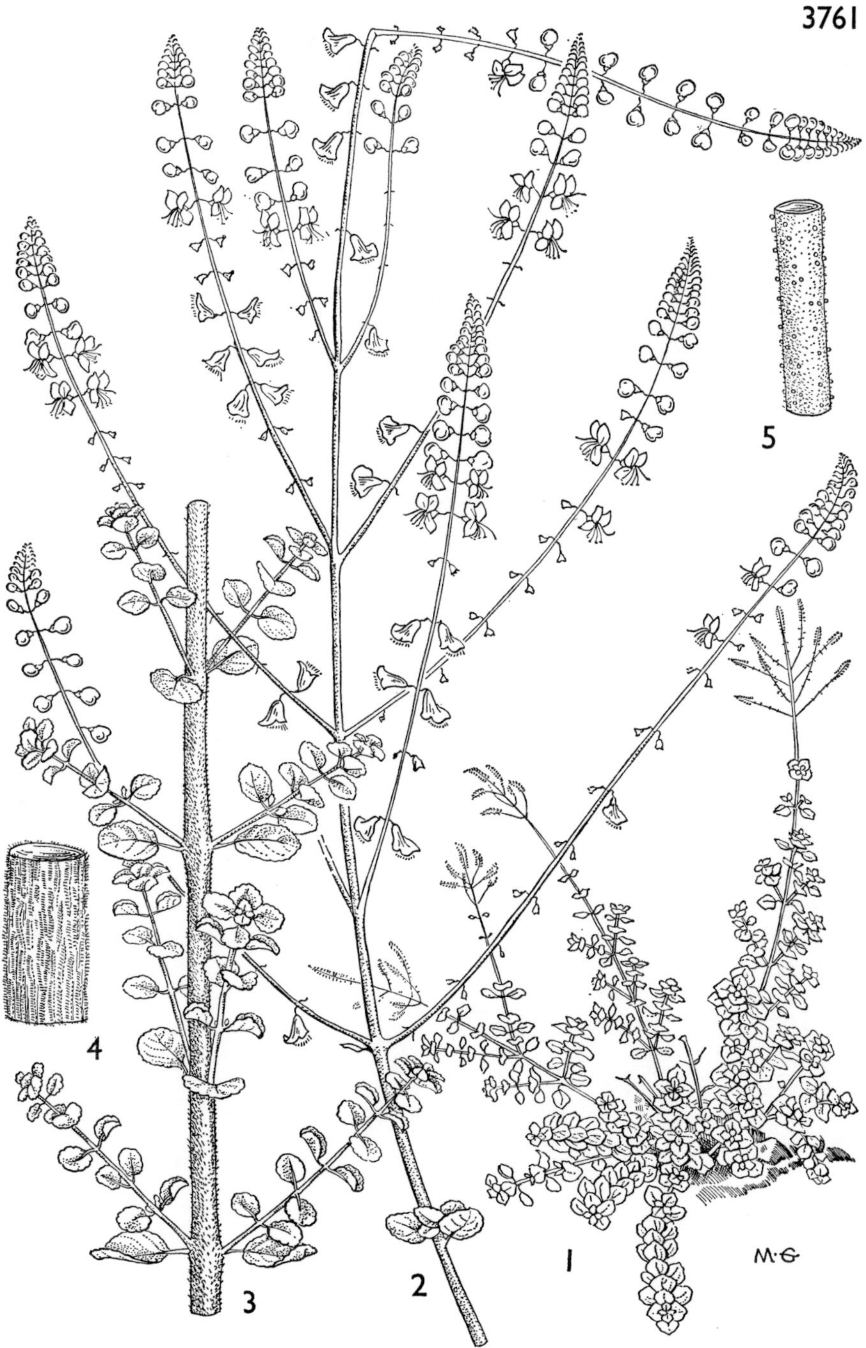

Eriope parvifolia Mart. ex Benth. subsp. *glandulosa* Harley

3762

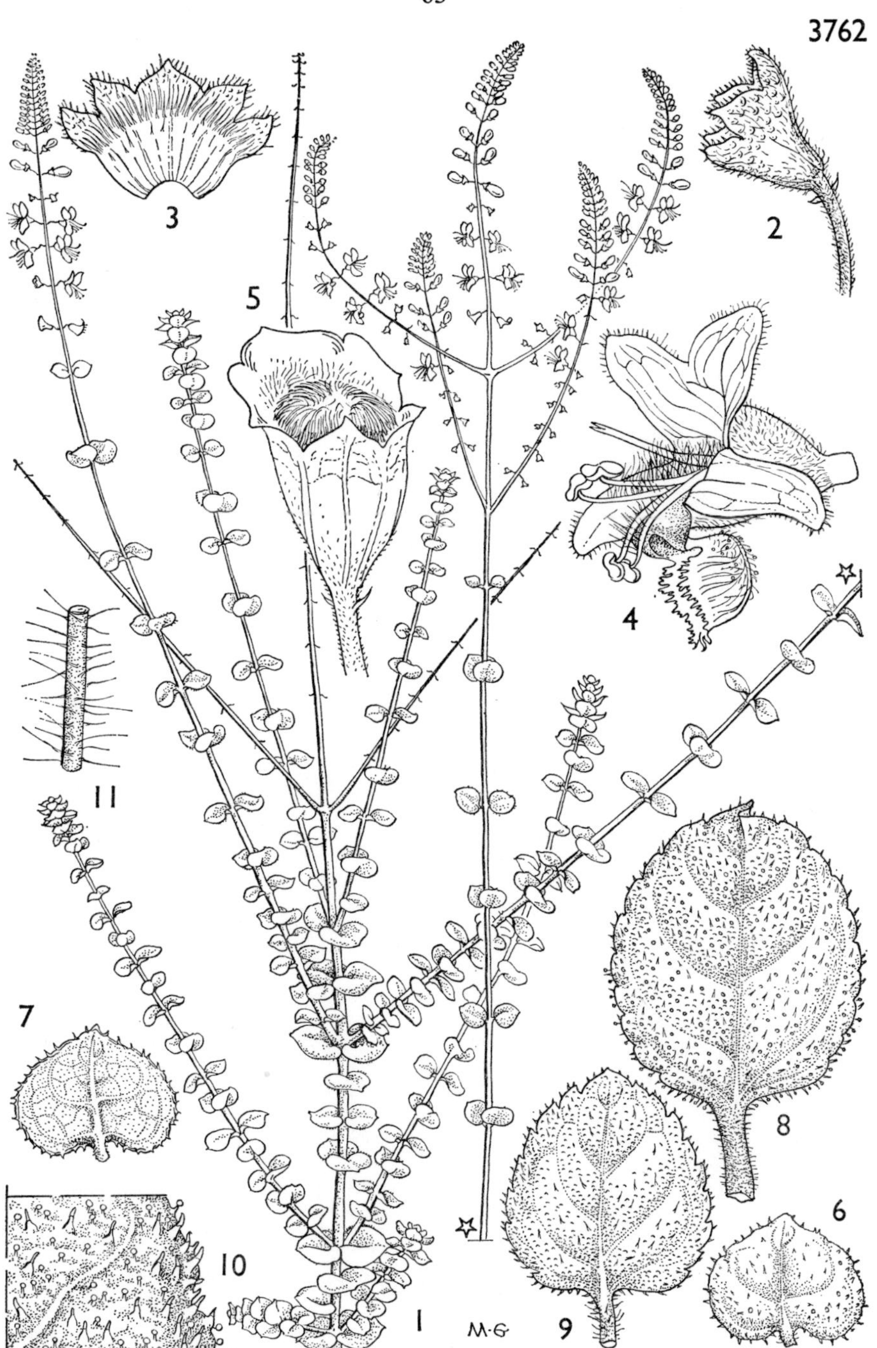

Eriope parvifolia Mart. ex Benth. subsp. *glandulosa* Harley (aff.)

of Lagoinha (5·5 km. SW of Delfino) on side road to Minas do Mimoso, c. 41°20′ W, 10°22′ S, in cerrado on sandstone rock exposures, alt. 950 m., 4. Mar. 1974, *Harley et al.* 16671 (CEPLAC!, K!, MO!, NY!, P!, RB!, U!, UCLA!, US!).

This attractive plant was found in some profusion growing in rather drier areas along the roadside. Its affinities are not very clear. In indumentum it recalls *E. macrostachya* though there seems to be a complete absence of stalked glands or glandular hairs, and the possession of very striking fistulose swellings below the inflorescence is also found in *E. hypenioides* which grows in the same area. The locality is an isolated upland region, apparently not visited previously by botanists, in northern Bahia. Our collections indicate that it contains a wealth of new taxa.

TAB. 3763. *Eriope tumidicaulis.* FIG. 1, flowering shoot, × ⅔; 2, calyx, side view, × 8; 3, calyx, inner surface, × 8; 4, corolla, × 8; 5, leaf, lower surface, natural size; 6, leaf, upper surface, natural size; 7, fruiting calyx, side view, × 8; 8, nutlet, dorsal view, × 10. (All from *Harley et al.* 16671, type material.)

(*See page 67*)

11. **Eriope filifolia** *Benth.* in DC. Prod. 12: 141 (1848). Type: Brazil, Minas Gerais, Diamantina, Tejuco, *Riedel* 1166 (K!, lectotype; UCLA!, US!, photos; BR!, LE!, NY!, OXF!, US!, isotypes).
E. angustifolia Epling in Bull. Torr. Bot. Club 71: 494 (1944), *synon. nov.* Type: Brazil, Minas Gerais, Serra do Cipó, *Mello Barreto* 9124 (F!, holotype; UCLA!, photo).

Subshrub to 1·5 m., apparently often much less, with numerous slender branches. Lower parts of the stems, and the nodes above, tomentose with very short spreading hairs and stalked glands, and with scattered long spreading setae; upper part of the stems predominantly glabrous with glaucous bloom, internodes usually inflated. Lamina (25–)30–50(–80) mm. long, (1–)2–4(–5) mm. wide, narrowly linear and often curled, gradually tapering at apex, and at base imperceptibly into the short 1–2 mm. long petiole; upper surface densely glandular, and often with short spreading hairs, lower surface glabrous or almost so; margin incurved with usually obscure teeth. Inflorescence 20–40 cm. long, a simple raceme, or once-branched below, often densely glandular and hairy above. Calyx 2–2·5/1·75–2 mm. long, purple tinged, with appressed hairs and numerous stalked glands, shortly campanulate, two-lipped, becoming 3–(6–7) mm. in fruit, widely turbinate; pedicel 2–(4–5) mm., becoming 3–(6–7) mm. in fruit. Corolla c. 5·5 mm. long, violet. Anthers 0·75–1 mm. Nutlets 2–2·25 mm. long.

BRAZIL. Minas Gerais: Tejuco, Diamantina, Dec. 1824, *Riedel* 1166 (K!, LE!, NY!); Riacho das Varas, nr. Diamantina, 1892, *Schwacke* 8074 (G!); Serra do Cipó, Municipio de Santa Luzia, 6 Aug. 1936, *Archer & Barreto* 4965 (UCLA!, US!); 20 Sept. 1937, *Barreto* 9124 (F!); Serra do

Cipó, Chapeu do Sol, Dec. 1958, *A. P. Duarte* 4562 (HB!); Serra do Cipó, alt. 1150 m., 17 Feb. 1968, *Irwin, Maxwell, Wasshausen et al.* 20372 (NY!); Serra do Cipó, Municipio de Jaboticatuba, 5 Aug. 1972, *Hatschbach* 29947 (K!).

This species is at present known only from the Serra do Cipó and the Diamantina district, Minas Gerais, Brazil. Material in the Munich Herbarium, collected by Martius, with the details; 'campis editis siccis intus flum. Jequetinonha et Serra de Sant. Antonio, Serra Frio, Minas Gerais' is not this species, and is not an *Eriope*.

Specimens from Serra do Cipó differ slightly from the type in having broader, more hairy leaves, and longer pedicels.

TAB. 3764. *Eriope filifolia*. FIG. 1, flowering stem, × ⅔; 2, flowering stem, × ⅔; 3, flowering stem, × ⅔; 4, leaf, upper view, natural size; 5, leaf, *t.s.*, × 16; 6, corolla, side view, × 6; 7, calyx, side view, × 6; 8, calyx, inner surface, × 6; 9, indumentum from outer surface of lower corolla-lobe, × 26. (1 from *Riedel* 1166, lectotype; 2 from *Mello Barreto* 9124, holotype of *E. angustifolia* Epling; 3–9 from *Irwin et al.* 20372.)

(*See page 69*)

12. **Eriope hypenioides** *Mart. ex Benth.*, Lab. Gen. et Sp.: 142–3 (1833). Type: Brazil, Bahia, Rio das Contas, *Martius* 1903 (pro parte), (M! lectotype; UCLA!, US!, photos; M!, isotype).

Shrub or subshrub to 2 m., with the habit of *Hyptis* Sect. *Hypenia*. Lower part of stems dull brown, shortly pubescent and with glandular hairs, long spreading setae present, usually in localised patches; stems in the upper part glabrous except at the nodes, with glaucous bloom, deep purple-tinged, internodes weakly inflated. Lamina (20–)35–70(–80) mm. long, (14–)20–30(–34) mm. wide, ovate to ovate- or elliptic-lanceolate, coriaceous, with a more or less acute apex, and a truncate to weakly cordate base; lamina of the lowest leaves sometimes almost rotund; upper surface thinly hairy, soon becoming subglabrous; lower surface closely grey- to white-tomentose, particularly on the more or less incrassate veins; margin serrulate to serrate-dentate, not ciliate; petiole (2–)5–11(–13) mm. long. Inflorescence to 50 cm. long, once-branched, or lower branches branched again, branches glabrous with glaucous bloom, purple-tinged. Calyx 2·5–3/2–3 mm. long, turbinate, with appressed hairs, becoming 6–6·5/4–4·5 mm. long in fruit, two-lipped; pedicel 2–3·5 mm. long, becoming 3–4·5 mm. in fruit. Corolla 7–8 mm. long, pale to dark violet (pink in *Irwin, Harley, Smith et al.* 32553), with white markings in the throat at base of upper lobes and with pattern of dark blue veins. Anthers c. 1 mm. Nutlets 2–2·5 mm. long. $n = 10$, $2n = 20$.

BRAZIL. Bahia: Rio das Contas, ?1818, *Martius* 1903 (M!); Casa de Pedra, 1914, *Lützelburg* 30 (M!, W!); Carrasco, Rio Brumado, 1914, *Lützelburg* 147 (M!); entre Palmeiras e Lençois, alt. 900 m., 14 Sept. 1956, *E. Pereira* 2188 (RB!); summit of Morro do Chapeu, c. 7 km. S of the town of Morro

3763

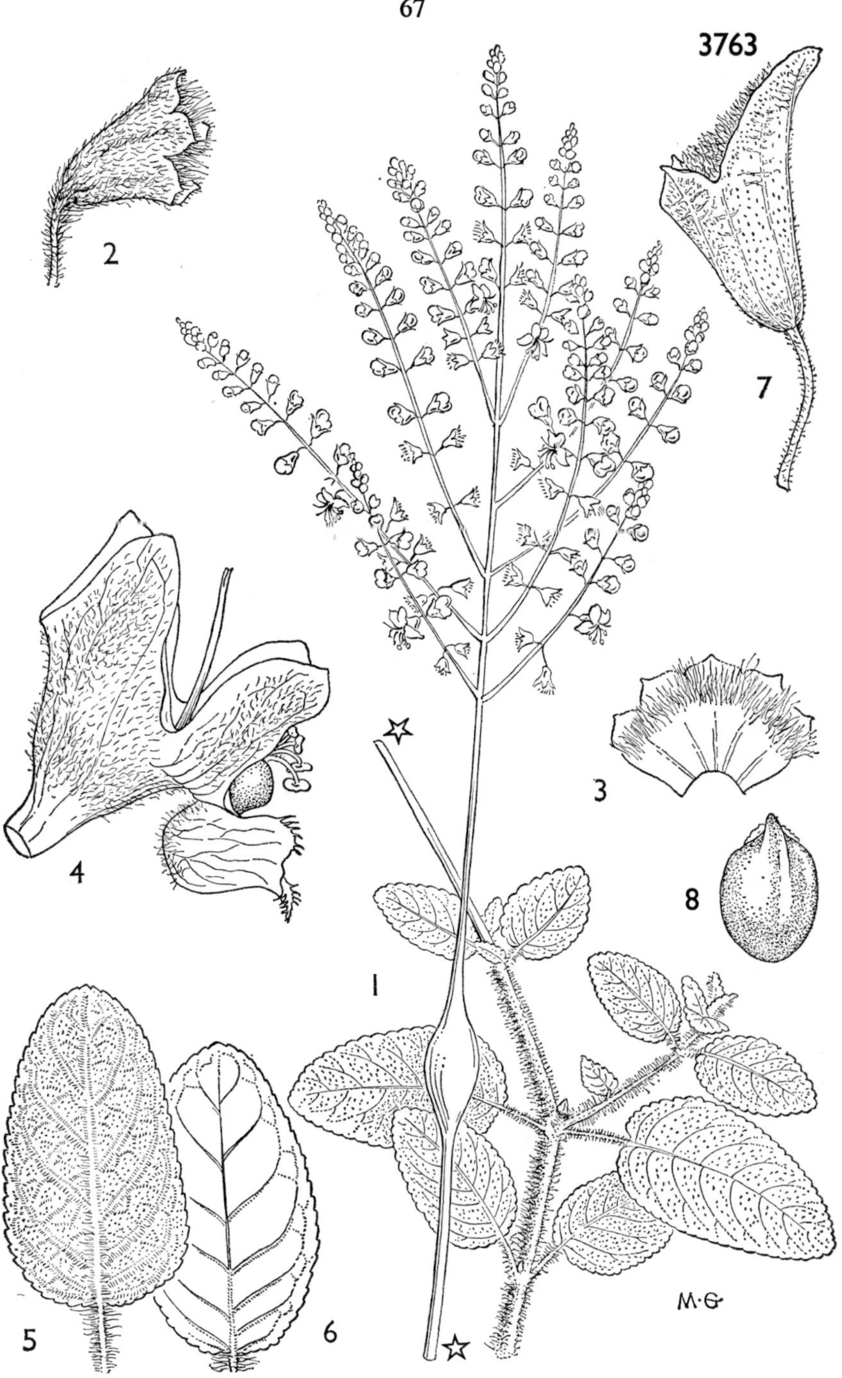

Eriope tumidicaulis Harley

Eriope filifolia Benth.

do Chapeu, sandstone with sand-filled depressions and crevices, alt. c. 1150 m., 16 Feb. 1971, *Irwin, Harley, Smith et al.* 32257* (K!, NY!, UB!) 32393 (K!, NY!, UB!); cerrado and low woodland, c. 5 km. S of town of Morro do Chapeu, near base of Morro do Chapeu, alt. c. 1100 m., 19 Feb. 1971, *Irwin, Harley, Smith et al.* 32553 (K!, NY!, UB!); 32554 (K!, NY!, UB!).

The type material at Munich is represented by four sheets of which one was designated as the lectotype by Epling in 1936. Unfortunately this appears to be a mixed sheet, consisting of three fragments. The two outer fragments are closely matched by those on the three other sheets, and are here selected to represent the type. The largest, central fragment apparently represents an as yet undescribed taxon within the genus. Though superficially rather similar, it differs in its weakly tetragonous stems, its larger leaves which are softly grey-velutinous above, and in the calyx which is larger, 4–4·5 mm. long in flower, and with five subequal, deltoid, acute teeth. The calyx-throat is also much less hairy. Thus in floral characters the fragment shows a relationship with *E. exaltata*, though clearly distinct from this. Further material is needed.

TAB. 3765. *Eriope hypenioides*. FIG. 1, habit, × $\frac{1}{18}$ approx.; 2, flowering shoot, × $\frac{2}{3}$; 3, leaf, from middle of stem, × $\frac{2}{3}$; 4, leaf, from lower part of stem, × $\frac{2}{3}$; 5, calyx, side view, × 6; 6, calyx, inner surface, × 6; 7, corolla, × 4; 8, nutlet, dorsal view, ×6. (All from *Irwin et al.* 32553, except for 3 from *Pereira* 2188, and 4 from *Irwin et al.* 32554.)

(*See page 73*)

13. **Eriope monticola** *Mart. ex Benth.*, Lab. Gen. et Sp.: 143 (1833). Type: Brazil, Bahia, in campis subalpestribus ad Sincorá, *Martius* 1947 (M!, holotype; UCLA!, US!, photos).

Subshrub to c. 1 m. Lower parts of stems densely grey-tomentose with short spreading hairs and few scattered long setae; stems in the upper parts glabrous except at the nodes, with glaucous bloom, deep purple-tinged; internodes not inflated (in only available material). Lamina 18–27 mm. long, 13–21 mm. wide, broadly ovate, strongly coriaceous, with a rather obtuse apex, often with a rather short apiculus, and with a rounded base; upper surface thinly hairy at first, rapidly becoming subglabrous and shining, but with persistent appressed hairs along the lower half of the midrib; lower surface appressed grey-tomentose with incrassate veins, margin lightly serrate-dentate, ciliate; petiole 3–4 mm. long. Inflorescence c. 20 cm. long, once-branched, panicle-branches pubescent with short grey hairs. Calyx 2/1·5–2 mm. long, turbinate, with appressed hairs, becoming 5/4 mm. long in (?immature) fruit, two-lipped, upper lip obscurely lobed, lower lip of two deltoid teeth. Pedicel 2 mm. long. Corolla c. 7 mm. long, pale violet. Anthers c. 1 mm.. Nutlets unknown.

---

* Collection from which chromosome numbers obtained.

Brazil, Bahia: Sincorá, October 1818, *Martius* 1947 (M!).

A species closely related to *E. hypenioides*, *E. polyphylla* and *E. crassifolia*.

TAB. 3766. *Eriope monticola*. FIG. 1, flowering shoot, × ⅔; 2, calyx, side view, × 8; 3, calyx, outer surface, × 8; 4, calyx inner surface, × 8; 5, corolla, side view, × 8; 6, leaf, upper view, × 2; 7, portion of leaf, upper surface, × 6; 8, portion of leaf, lower surface, × 6. (All from *Martius* 1947, holotype.)

(*See page 75*)

14. **Eriope crassifolia** *Mart. ex Benth.*, Lab. Gen. et Sp.: 143–4 (1833). Type: Brazil, Bahia, in Serra das Lages et Sincorá, *Martius* 1990 (M!, holotype; M!, isotype; UCLA!, US!, photos).

Subshrub to c. 60 cm., with leaves densely imbricate and crowded at the ends of the rather stout branches. Bare stems longitudinally striate in the lower parts, and with conspicuous leaf-abscission scars above, hairy with short spreading or upwardly directed hairs and very sparse long spreading setae. Leaves subsessile; lamina 9–11 mm. long, 4–9 mm. wide, broadly to narrowly ovate, thick and rigid, with broadly acuminate apex and rounded base; subglabrous on both surfaces but with sparse pilose hairs on the veins, midrib and primary veins prominent below; margin serrate-dentate with rather few prominent acute teeth, glabrous. Inflorescence up to 14 cm. long, a simple raceme (in available material), with indumentum as on stem, but denser and mixed with glandular hairs. Calyx c. 3–3·5 mm. long, turbinate, weakly two-lipped, with spreading non-glandular and glandular hairs, becoming 7/6 mm. long in fruit, with persistent spreading hairs, teeth of upper lip remaining distinct; pedicel 1·5–2 mm. becoming 2–3 mm. long in fruit. Corolla 'dilute violaceous' (no longer extant in herbarium material). Nutlets c. 3 mm. long.

Brazil, Bahia: Serra das Lages et Sincorá, November 1818, *Martius* 1990 (M!).

Closely related to *E. polyphylla*.

TAB. 3767. *Eriope crassifolia*. FIG. 1, flowering shoot, × ⅔; 2, leaf, upper view, × 2; 3, leaf, lower view, × 2; 4, leaf, upper view, × 2; 5, calyx, × 9; 6, calyx, inner surface, × 9; 7, fruiting calyx, side view, × 9. (All from *Martius* 1990, holotype.)

(*See page 77*)

15. **Eriope polyphylla** *Mart. ex Benth.*, Lab. Gen. et Sp.: 143 (1833). Type: Brazil, Bahia, in campis subalpestribus ad Rio de Contas et Lages, *Martius* s.n. (M!, holotype; UCLA!, US!, photos).

E. obtusata Benth., Lab. Gen. et Sp.: 143 (1833), *synon. nov.* Type: Brazil, Bahia, in campis subalpestribus ad Rio de Contas et Lages, cum *E. polyphylla*, *Martius* s.n. (M!, holotype; UCLA!, US!, photos).

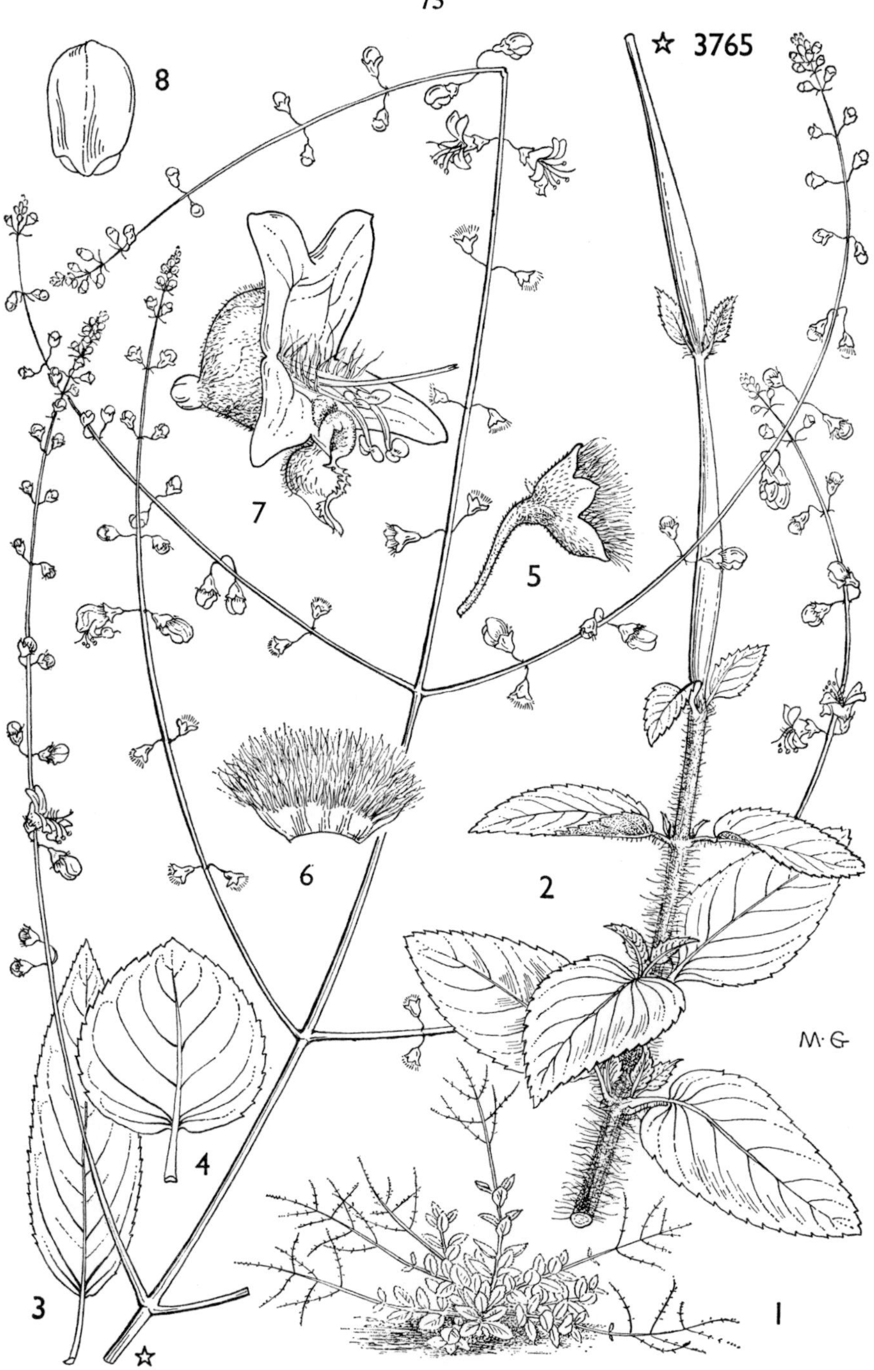

Eriope hypenioides Mart. ex Benth.

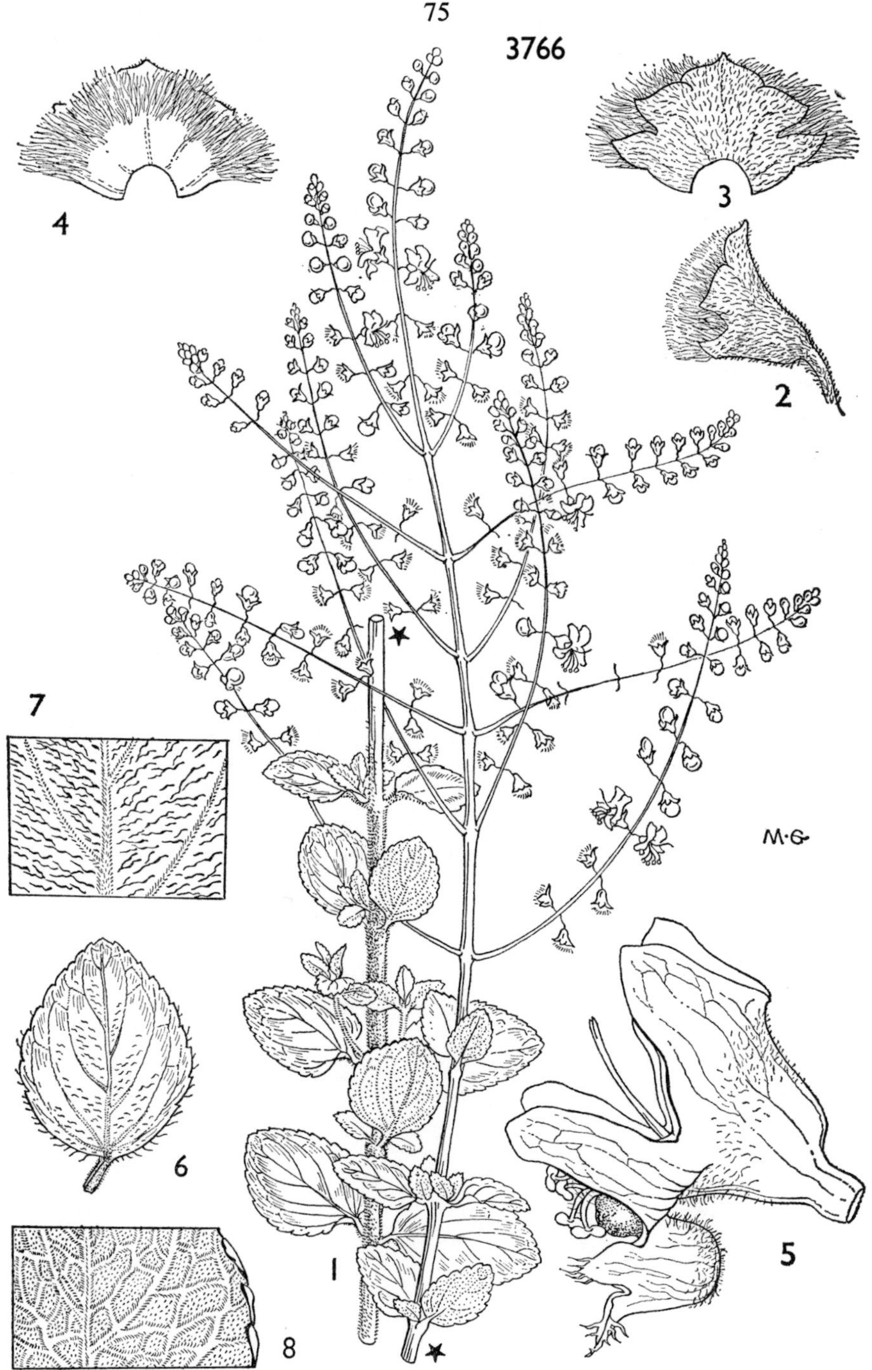

Eriope monticola Mart. ex Benth.

3767

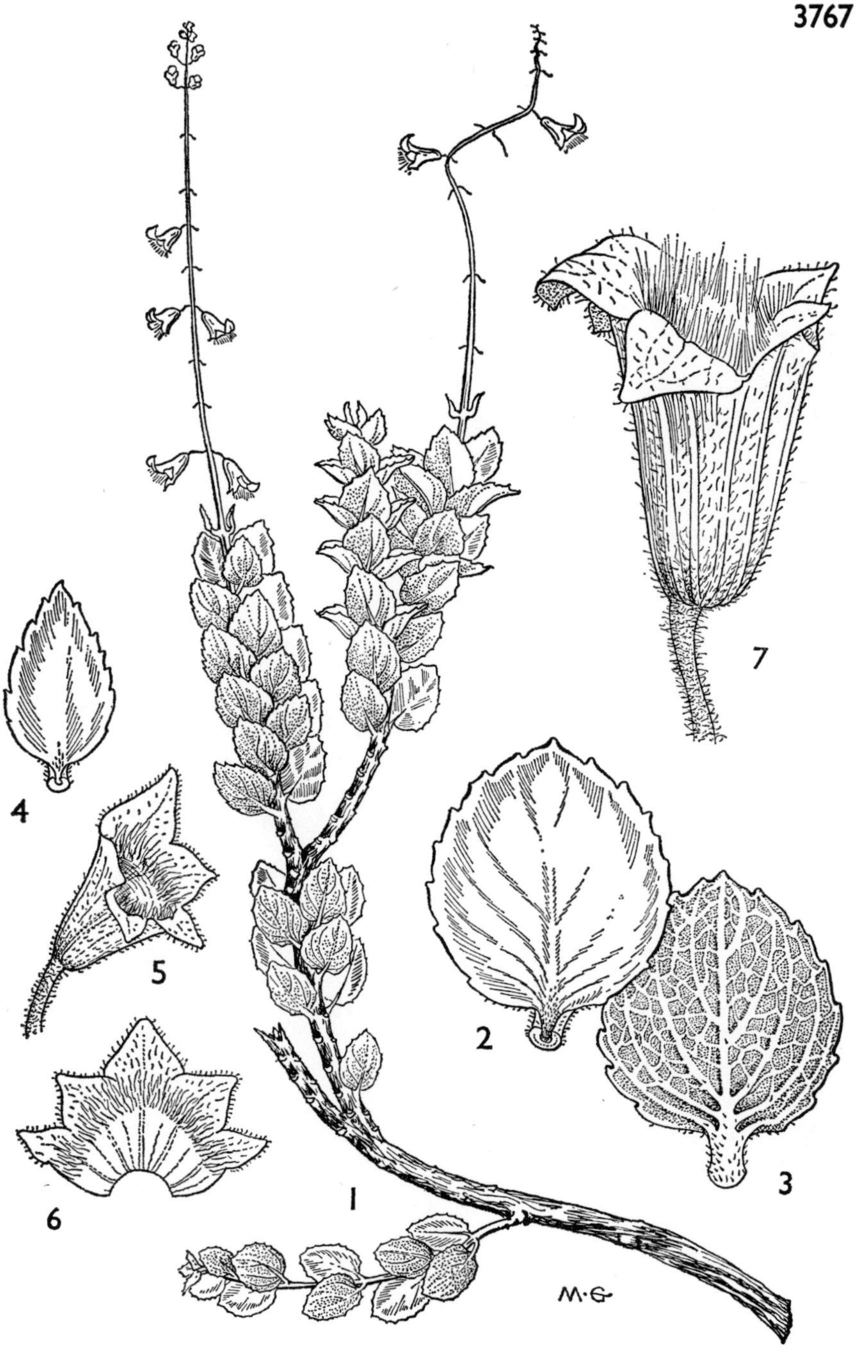

Eriope crassifolia Mart. ex Benth.

Much-branched subshrub of unknown height, with stems dark brown and sparsely hairy in the lower part, becoming densely tomentose above, with short spreading or upwardly directed hairs, and with long spreading setae scattered among the leaves; internodes not inflated. Lamina (10–)12–14 mm. long, 4–8 mm. wide, ovate-elliptic (often narrowly so), thick and coriaceous, with rounded base and obtuse to acute apex; upper surface more or less subglabrous, but with a few short appressed hairs along lower ⅓ of midrib, and scattered near margin, lower surface appressed-tomentose, with prominent midrib and primary veins; margin long-ciliate, especially in the lower half, obscurely toothed with few teeth; petiole up to 2 mm., usually less. Inflorescence to 22 cm., usually shorter, a simple raceme, or with few short branches below, with indumentum as on the stem but paler and denser and mixed with glandular hairs. Calyx 2·5–3/2–2·5 mm. long, turbinate, two-lipped with whitish spreading hairs mixed with glandular hairs especially on the tube, becoming 5–6/4–4·5 mm. long in fruit, with persistent spreading hairs. Pedicel 2 mm. becoming 2·5–3 mm. long in fruit. Corolla unknown. Nutlets (?ripe) 2 mm. long.

BRAZIL, Bahia: Rio de Contas & Lages, Oct. 1818, *Martius* s.n. (M!).

According to Bentham, the specimens which he described as *E. obtusata* and *E. polyphylla* were originally together as one gathering in Martius's Brazilian herbarium. Bentham considered them distinct, however, the former having obtuser, more sessile and more tomentose leaves and more shortly branched, more villous racemes. It is interesting to note that both specimens were annotated by Epling prior to his account of 1936. Although in this work he treats the two as distinct species, his earlier annotations show that he believed the two to be conspecific. From the scrappy material available, there seems little reason to doubt that this earlier view is correct, and that the variation displayed is no more than one would expect to find in a single population.

TAB. 3768. *Eriope polyphylla*. A, fig. 1, flowering shoot, × ⅔; 2, leaf, upper view, × 3; 3, leaf, lower view, × 3. B, fig. 4, flowering shoot, × ⅔; 5, leaf, upper view, × 3; 6, leaf, lower view, × 3. (A: 1, 2 & 3 from *Martius* s.n., holotype of *E. obtusata* Benth. which is conspecific with B: 4, 5 & 6 from *Martius* s.n., holotype of *E. polyphylla* Mart. ex Benth.)

(*See page 81*)

16. **Eriope crassipes** *Benth.*, Lab. Gen. et Sp.: 144 (1833). Types: Brazil, Goiás, paturages récemment brulés près l'aldea de St. Jozé, partie méridionale de la province, *St.-Hilaire* 781 (P!, lectotype); in provincia Rio de Janeiro, *Sello* s.n. (?B†, not located).

a. subsp. **crassipes**

E. nudiflora Kunth ex Benth., Lab. Gen. et Sp.: 144 (1833). Type: Colombia, Maipurés, *Humboldt & Bonpland* s.n. (B†, holotype; F!, K!, UCLA!, US?, photos).

E. teucrioides St.-Hil. ex Benth., Lab. Gen. et Sp.: 144 (1833), *synon. nov.*
Types: Brazil, Minas Gerais, carrascos (forêts naines) nouvellement brulés
près la fazenda (habitation) de José Caetano de Mello, dans les Minas
Novas, *St.-Hilaire* 1662 (P!, lectotype); Minas Gerais, plateau découvert
entre Alto dos Bois & Villa do fanado (Minas Novas), *St.-Hilaire* 1364 (P!,
paratype; UCLA!, photo).

E. chamaedrifolia Taub. ex Glaziou in Mém. Soc. Bot. Fr. 3: 556 (1911),
nomen nudum. Cited specimen: Brazil, Minas Gerais, Arassuahy, dans le
campo, 1883–4, *Glaziou* 15313 (G!, K!, LE!, P!, R!).

E. crassipes Benth. var. *acutifolia* Benth., Lab. Gen. et Sp.: 144 (1833) and,
for first citation of types, in DC. Prod. 12: 142 (1848). Types: Brazil,
Minas Gerais, prope fazenda da San Rita, *Sello* 1502 (B†, UCLA!, lectotype; F!, K!, UCLA!, US!, photos; K!, W!, isotypes); Brazil, Goiás (or possibly Bahia), dry bushy, sandy campos, Mission of Duro, Oct. 1839, *Gardner*
3387 (BM!, BR!, G!, K!, NY!, OXF!, P!, SP!, UCLA!, W!, paratypes).

E. crassipes Benth. var. *parvifolia* Benth. in DC. Prod. 12: 142 (1848). Type:
Brazil, Goiás, Chapada da Mangabeira, Sept. 1839, *Gardner* 3386 (K!,
holotype; US!, photo; BM!, BR!, K!, OXF!, isotypes).

E. crassipes Benth. var. *macrophylla* Benth. in DC. Prod. 12: 142 (1848).
Type: Brazil, São Pulo [*sic*], *Pohl* 580 (K!, holotype; US!, photo; UCLA!,
W!, isotypes).

Usually erect, musky-scented herb with several often unbranched stems to
50(–80) cm. high from a woody often digitately lobed swollen rootstock.
Stems with short more or less crisped hairs, mixed with glandular hairs above,
and usually with numerous long spreading setae in the lower parts. Lamina
(30–)40–60(–75) mm. long, (10–)20–30(–35) mm. wide, rarely smaller, broadly
to narrowly elliptic-ovate or ovate-lanceolate, often coriaceous, with acute,
rarely rounded and obtuse, apex, and weakly cordate to attenuate base;
both surfaces usually sparsely hairy at first, becoming nearly glabrous except
for appressed hairs along the sunken midrib above (hairier in some populations in Minas Gerais & Bahia); margin weakly ciliate, usually serrate to
serrate-dentate with numerous irregular often salient or occasionally obscure
teeth; petiole (1–)3–5(–15) mm. long, usually conspicuously jointed near the
base. Inflorescence 10–30 cm. long, once-branched with few branches, the
lowest often the longest, rarely simple; indumentum as on the upper stems,
glandular-hairy. Calyx 2–3/1·5–2·5 mm. long, turbinate, with appressed
hairs and scattered short-stalked glands, becoming (5–)6–8/5–6(–7) mm. long
in fruit, narrowly campanulate-turbinate. Pedicel 1·5–2·5 mm., becoming 2–3
mm. long in fruit. Corolla 6–8 mm. long, violet, with paler area at base of upper
lip, overlaid with darker lines. Anthers c. 1 mm. Nutlets 2·5–3 mm. long.

FRENCH GUIANA: Village Ouanari, 10 Nov. 1954, *Black, Vincent & Colmet
d'Aage* 54-17626 (UNB!).

BRAZIL, Pará: Belém, Rio Guajará, campina, 14 May 1954, *Black* 54-16187
(UB!); Serra do Ariranha, Rio Trombetas, mata baixa, em terreno pedregoso,

3768

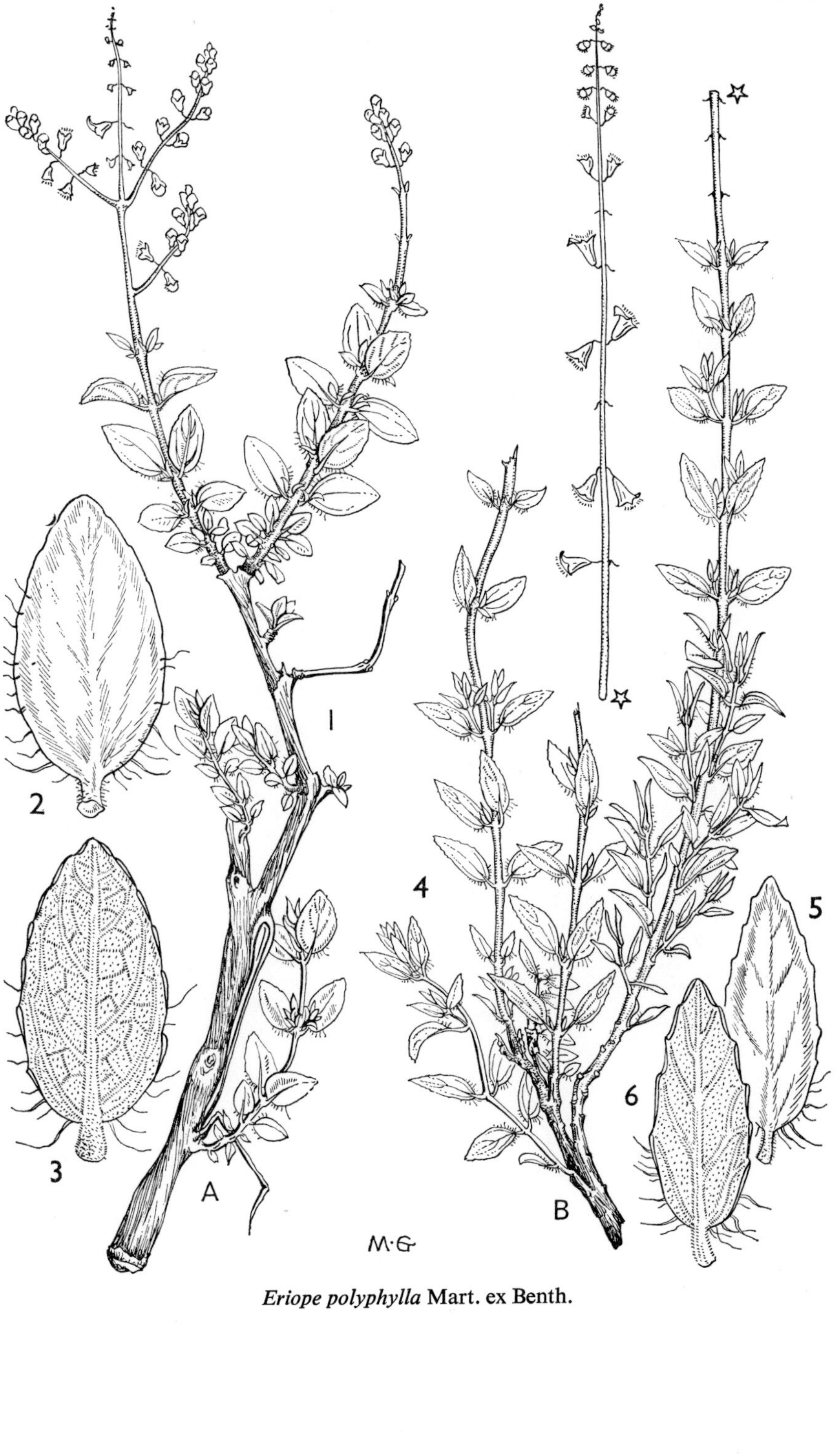

Eriope polyphylla Mart. ex Benth.

83

12 Dec. 1910, *Ducke* 954 (RB!); Serra do Cachimbo, 20 Sept. 1955, *Pereira* 1870 (F!, RB!, UCLA!); Serra do Cachimbo, alt. 425 m., 14 Dec. 1956, *Pires, Black, Wurdack & Silva* 6252 (NYBG!).

Mato Grosso: Cuiabá, 30 Nov. 1893, *Malme* s.n. (S!); Burity, NE of Cuiabá, in open campo, June 1927, *Dorrien Smith* 305 (K!); S. Anna da Chapada, 20 July 1902, *Robert* 402 (BM!); 1 Sept. 1902, 540b (K!, BM!); 9 Oct. 1902, 606 (K!, BM!); S. Anna da Chapada, 23 July 1902, *Malme* 2090 (S!); Braço Rio Arinos, 26 Sept. 1943, *Baldwin* 3057, 3080 (US!); Rio Grande, 9 Oct. 1938, *Rombouts* 2738 (SP!); Estaça, Campo Grande, 10 Sept. 1936, *Archer & Gehrt* 157 (SP!, US!); ca. 15 m. S of Garapú, 4 Oct. 1964, *Irwin & Soderstrom* 6653 (NY!); Corrego das Moreiras, Sept. 1914, *Kuhlmann* 1311 (SP!), *Hoehne* 7341 (UCLA!); Xavantina, 11 Oct. 1967, 12 Oct. 1967. *Argent et al.* 6712 (K!), 6724 (K!); Xavantina-Cachimbo Expedition Base Camp, 12°49′ S, 51°46′ W, 6 Aug. 1968, *Richards* 6605 (K!); 16 Sept. 1968, *Harley et al.* 10056 (K!); 22 Sept. 1968, 10183 (K!); 28 Sept. 1968, 10319 (K!); 5 Nov. 1968, 10948 (K!); 6 Oct. 1968, *Fonseca & Onishi* 1186–47, (K!, UB!); 6 Oct. 1968, 1145-366, (K!, UB!); 11 Oct. 1968, 1411-632 (UB!).

Pernambuco: Inajá. Margem da estrada de Ibimirim a Petrolandia, 13 July 1954, *Andrade-Lima* 54-1978 (IPA!, K!).

Goiás: Mission of Duro, dry bushy sandy campos, Oct. 1839, *Gardner* 3387 (BM!, BR!, G!, K!, NY!, OXF!, P!, SP!, UCLA!, W!); Serra do Pireneos, Aug. 1892, *Ule* 754 (R!); Perinópolis, 22 July 1952, *de Oliveira* 5, *Macedo* 3606 (NY!, UCLA!, US!); Municipio Anapolis, Estação experimental, 17 Oct. 1956, *Smith & Macedo* 4756 (US!); Cavalcante, 30 Sept. 1829, *Burchell* 7922 (K!); Chapada da Mangabeira, Sept. 1839, *Gardner* 3386 (BM!, BR!, K!, OXF!); Caiapónia to Aragarças, a 9 km. de Caiapónia, 3 Oct. 1968, *Onishi & Fonseca* 271-1040 (K!); c. 33 km. S of Caiapónia on road to Jataí, 19 Oct. 1964, *Irwin et al.* 7074 (NY!).

Brasília, Distrito Federal: Campus de Universidade, 29 May 1965, *Sucre* 411 (UB!); Horto do Guará, 22 Aug. 1961, *Heringer* 8604-798 (UB!); Catetinho, cerrado, 12 Apr. 1963, *Pires et al.* 9034 (UB!); Chapada da Contagem, c. 20 km. E of Brasília, 1000 m. alt., 14 Oct. 1965, *Irwin et al.* 9240 (K!).

Bahia: c. 8 km. NW of Barreiras, by incomplete road to Santa Rita da Cassia, cerrado on slopes of Espigão do Mestre, among boulders, 3 Mar. 1972, *Irwin, Harley, Smith et al.* 31423 (K!, NY!, UB!); c. 80 km. W of Barreiras, cerrado, 7 Mar. 1971, *Irwin, Harley, Smith et al.* 31663 (K!, NY!, UB!).

Minas Gerais: Municipio Gouveia, Serra do Espinhaço, 6 Sept. 1971, *Hatschbach* 27319 (K!); Lagoa Santa, 5 Nov. 1963, *Warming* 932 (P!); Caldas, *Widgren* 425 (BR!, K!, LE!, S!); *Regnell* 1-318 (F!, K!, MO!, R!, S!); 20 Sept. 1875 *Mosen* 3996 (S!, UCLA!); Cipó, ao Rio Preto, 26 Apr. 1892, *Glaziou* 19685 (K!, R!); margem do Paraopeba, Baxia de Tres Marias, Felixlandia, 26 Sept. 1959, 20 Jun. 1959, *Heringer* 7100 (UB!). 7035 (UB!); Fazenda do Chico Mauriso, 25 Apr. 1957, *Heringer* 5845 (UB!); Arassuahy, dans le campo, 1883–84, *Glaziou* 15313 (G!, K!, LE!, P!, R!); Belo Horizonte, *Damazio* s.n. (K!); 25 km. SW of Belo Horizonte, above village of Barreiras, alt. 1200 m., 25 July 1944, *Williams & Assis* 8172 (R!, UCLA!); près la fazenda de Jozé Caetano de Mello, Minas Novas, 1816–1821, *St.-Hilaire* 1662 (P!); plateau découvert entre Alto dos Bois & 'Villa do fanado' (Minas Novas). *St.-Hilaire* 1364 (P!).

Rio de Janeiro: in provincia Rio de Janeiro, *Sello* s.n. (?B†).

São Paulo: São Paulo (capital), 11 Dec. 1945, *Pickel* 2434 (UCLA!); Fazenda Campo Grande, Campinas, 4 Dec. 1938, *Viegas et al.* 3155 (SP!); Itapetinanga, campo do mato seco, 25 Sept. 1887, *Löfgren* 195 (R!, SP!); Itapetinanga, 13 Nov. 1946, *Iglesias de Lima* s.n. (RB!); 1 Oct. 1959, *Machado de Campos* 63 (SP!); Bom Retiro, SW da cidade de Itapetinanga, 25 Nov. 1959, *Machado de Campos* 135 (NY!, US!); Santa Rita do Passa Quatro, 25 Oct. 1897, *Hemmendorff* 30 (MO!, S!); Municipio de Itirapina, 3·1 km. along road east from the RR at Itirapina, 22°15·3′ S, 47°47·2′ W, alt. 800 m., 30 Nov. 1961, *Eiten & Freitas Campos* 3404 (US!); Itirapina, 5 km. da rodovia Washington Luiz na estrada Itirapina-Rio Claro, 13 Sept. 1962, *Felipe* 82 (RB!, US!); São José dos Campos, 2 Feb. 1909, *Löfgren* 44 (S!); 31 May 1961, *Eiten & Sendulsky* 2866 (NY!); 7 km. SW of S. José dos Campos, 10 Aug. 1961, *Mimura* 16 (F!, G!, K!, NY!, US!); Ypiranga, 10 Oct. 1922, *Gehrt* 8054 (SP!); Municipio de Mogi-Guaçu, 10 km. NNW of Padua Sales, Fazenda Campininha, 18 Nov. 1960, 17 Nov. 1960, *Mattos & Mattos* 8543 (NY!), 8496 (US!); Mogi-Mirim, 15 Nov. 1901, *Hammar* 46 (SP! UCLA!); 15 Aug. 1827, *Burchell* 5182, (K!); São Bernardo, 26 Oct. 1913, *Brade* 6347 (SP!, UCLA!); Botucatú, *Edwall* s.n. (SP!, UCLA!); São Miguel Arcanjo, 16 Sept. 1959, *Machado de Campos* 5 (SP!, US!).

Paraná: Municipio Jaguariaiva, Barra Rio das Mortes, 24 Mar. 1968, *Hatschbach* 18951 (F!, MO!, NY!, US!); Jaguariaiva, in campo, 25 Oct. 1910, *Dusén* 10480 (K!, NY!, S!, US!); Municipio Sengés, Rio Velame, 17 June 1971, *Hatschbach* 26787 (K!); inter Sengés et Fabró Rego, 11 Dec. 1910, *Dusén* 10977 (BM!); Municipio Arapoti, Fazenda do Tigre, 9 Sept. 1960, *Hatschbach* s.n. (HB!); Itararé, opp. Morungava, alt. 760 m., 14 Feb. 1915, *Dusén* 16692 (MO!); same locality, alt. 720 m., 30 Jan. 1915, *Dusén* 16572 (MO! UCLA!); Morungava, 5 Dec. 1915, *Dusén* 17414 (S!).

Colombia: Cundinamarca, Nocaima, Hacienda Tobia, alt. 850 m., 15–20 Jan. 1942, *Garcia Barriga* 10583 (US!); Prov. of Bogotá, S. Martin, 1844, *Goudot* 1 (P!); llano de S. Martin, *Karsten* s.n. (LE!); plains de S. Martin, 1866, *Triana* 1093 (G!); llanos, alt. 1500 ft., Meta region, Mar. 1948, *Sandeman* 5869 (K!, US!); plains du Meta, Bogotá, alt. 300 m., 1851–57, *Triana* s.n. (P!); San José de Ocuné, Hacienda de Hector Perez, Saracuré on the Rio Vichada, c. 40 km. west, 18 Jan. 1944, *Hermann* 10958 (UCLA!); c. 24 km. NE of San José de Ocuné, dry sandy llanos bordering the Rio Vichada, alt. 100 m., 21 Jan. 1944, *Hermann* 11039 (UCLA!); Dept. Boyacá, Rio Guanapolo, 3 Mar. 1939, *Haught* 2655 (NY!, UCLA!, US!); Maypure (Maipurés), Apr. 1800, *Humboldt & Bonpland* 898 (B, photos in F!, K!, UCLA!, US!).

Venezuela: prope Maypures (Maipurés) ad flumen Orinoco, June 1854, *Spruce* 3618 (BR!, K!, UCLA!, W!); Barinas, nr. Santa Barbara de Barinas, Apr. 1942, *Lasser* 227 (US!).

Bolivia: Prov. de Chiquitos, de la cima delo cerro de Santiago, *d'Orbigny* 889 (P!); Prov. del Sara, Cant. Buena Vista, Santa Cruz, 28 Mar. 1916, *Steinbach* 2701 (F!), 8 Sep. 1924, 6405 (G!, K!, S!).

The very wide distribution of *Eriope crassipes* almost encompasses that of the entire genus. Not surprisingly there is a considerable amount of morphological variation, often more or less geographically correlated, and in the following two cases meriting subspecific rank.

Within subsp. *crassipes*, however, plants from the eastern part of its range, especially E. Minas Gerais, Bahia etc., seem to have a more diffuse habit with smaller, relatively broader, longer-petioled leaves thinly hairy above and pilose beneath, sometimes rather densely so. These plants have been treated as separate species: *E. teucrioides* Benth. and *E. chamaedrifolia* Taub., ined. Further field observations are necessary, but they appear to be local variants of the polymorphic and widespread *E. crassipes* and may require subspecific status. Of the two specimens cited by Bentham under *E. teucrioides*, Epling selected *St.-Hilaire* 1662 as the type, although erroneously annotating the other, *St.-Hilaire* 1364, as type on the sheet itself. This latter specimen is rather different, but inadequate for any decision to be made as to its status. It may merely represent a depauperate plant, although it differs not only in its smaller leaves, but in its smaller calyces also.

The collections from Maipurés (variously spelt Maypure and Maypures) are difficult to localise as it borders on both Venezuela and Colombia. I have followed Dugand (El primer ambo de Humboldt a la Nueva Granada, in Rev. Acad. Colomb. 9: 210–213, 1954) in assuming that Humboldt and Bonpland's gathering (the holotype of *E. nudiflora* Kunth ex Benth.) came from the Colombian bank of the river, while *Spruce* 3618 is generally accepted to be from the Venezuelan side.

I have taken the step of changing the lectotype of *E. crassipes*, as selected by Epling (in Fedde Rep. Spec. Nov. Beih. 85: 192, 1936). This was said to be *Sello* 1501, from Minas Gerais State, near Tapera, Brazil, in the herbarium of the Botanisches Museum, Berlin (B), now presumably destroyed. However, when *E. crassipes* was originally published. Bentham mentioned only one Sello specimen, said to be from Rio de Janeiro State. The material I have seen of *Sello* 1501, from Kew (K!), Paris (P!), Vienna (W!) and the herbarium of the University of California, Los Angeles (UCLA!) is unfortunately unlocalised and there is no mention of Tapera or Minas Gerais. I have therefore taken the St.-Hilaire specimen, cited first by Bentham, as the new lectotype.

The type of var. *acutifolia* Benth., which I have not taken up, was in Berlin, and is now presumed destroyed. However, photographs of it exist and from these it is clear that the small fragment in the Herbarium of the University of California, Los Angeles, identified as this, was actually removed from the Berlin sheet and therefore is holotype material.

TAB. 3769. *Eriope crassipes* subsp. *crassipes*. FIG. 1, flowering shoot, × ⅔; 2, basal part, showing swollen rootstock, × ⅔; 3, habit of small plant; 4, detail of node to show jointed petiole base, × 3; 5–9, leaf, upper surface, to show variation, × ⅔; 10, calyx, side view, × 6; 11, calyx, inner surface, × 6; 12, corolla, side view, × 6; 13, fruiting calyx, side view, × 6. (1 & 5 from *Hatschbach* 1895; 2 & 4 from *Harley et al.* 10056; 3 & 9 from *Dessys* 1549; 6, 7 & 13 from *Harley et al.* 10319; 8 from *Richards* 6605; 10–12 from *Irwin et al.* 31663.)

b. subsp. **trichopoda** (*Briq.*) *Harley* comb. & stat. nov.

E. trichopoda Briq. in Bull. Trav. Soc. Bot. Genève 5: 115–6 (1889). Type: Paraguay, Cordillère de Peribébuy, dans les clairières des forêts, *Balansa* 1011 (G!, holotype; BR!, K!, LE!, P!, isotypes).

Similar to subsp. *crassipes*, but stems in lower part densely covered with long spreading setae; stems in upper part glabrous, with glaucous bloom, and weakly inflated below the inflorescence. Lamina rather more hairy, ovate-elliptic.

PARAGUAY: Cordillère de Peribébuy, dans les clairières des forêts, 11 Jan. 1877, *Balansa* 1011 (BR!, G!, K!, LE!, P!); in valle fluminis Y-acá, *Hassler* 7002a (G!); in regione cursus superioris fluminis Y-acá, Central Cordillera, ? Jan. 1900, *Hassler* 6994 (F!, G!, K!, MO!, NY!, P!, RB!, S!, US!, W!).

These records appear to be all from SW Paraguay. Subsp. *crassipes* does not occur in Paraguay.

c. subsp. **cristalinae** *Harley*, subsp. nov., a typo caulibus florigeris glabris pruina ceracea purpureo-tinctis, internodiis proxime sub inflorescentia valde inflatis; a subsp. *trichopoda* caulibus setis longis patentibus carentibus, ab utraque foliis minus hirsutis magis carnosis marginibus cartilagineis, petiolis longioribus 6–10(–13) mm. longis differt.

Habit as in subsp. *crassipes*; stems to 50 cm. tall, with short crisped hairs in the lower part, apparently eglandular; long spreading setae absent. Lamina 40–60 mm. long, 14–26 mm. wide, ovate- to obovate-lanceolate, rather fleshy, with rounded to acute apex and cuneate to attenuate base, less hairy, often subglabrous except for appressed hairs along the sunken midrib above and scattered hairs on the midrib beneath. Margin cartilaginous, weakly ciliate. Petiole 6–10(–13) mm. long. Flowering stems as in subsp. *trichopoda*, glabrous, purple-tinged with a waxy bloom and strongly inflated internodes between the upper leaves and the lower part of the inflorescence.

BRAZIL, Goiás: Cristalina, planta com 50 cm., flores roxeadas, ramos surgindo após a queimada do cerrado, 15 Aug. 1967, *Heringer* 11539 (K!, holotype; NY!, UB!, isotypes); 5 km. W of Cristalina, frequent, cerrado, alt. 1175 m., subshrub [*sic*] c. 50 cm. tall, stems purplish-green, fruit brown, 2 Nov. 1965, *Irwin et al.* 9795 (K!).

3769

Eriope crassipes Benth. subsp. *crassipes*

89

These two gatherings indicate the presence of a distinct population near Cristalina. Their general appearance does not suggest a close relationship with subsp. *trichopoda*. The possession of the 'greasy pole' syndrome has no doubt been attained independently in each case through the action of natural selection on the genetic make-up of the neighbouring, typical, populations.

TAB. 3770. A: *Eriope crassipes* subsp. *cristalinae*, fig. 1, habit, × ⅓; 2, flowering shoot, × ⅔. B: *E. crassipes* subsp. *trichopoda*, fig. 3, habit, × ⅓; 4, flowering shoot, × ⅔; (A from *Irwin et al.* 9795; B from *Balansa* 1011, isotype.)

(*See page 91*)

17. **Eriope obovata** *Epling* in Fedde Rep. Spec. Nov. Beih. 85, 2: 191 (1936). Type: Brazil, Bahia, Serra do Sincorá, *Ule* 7348 (K!, holotype; UCLA!, photo; B†, isotype; F!, US!, photos).

Erect herb to c. 35 cm. high, with stem branched near the base, branches erect. Underground parts unknown. Stems with short spreading hairs in the lower part; long spreading setae absent. Lamina 18–28 mm. long, 8–10 mm. wide, obovate-lanceolate to elliptic, apparently fleshy, with rather obtuse apex and attenuate base; both surfaces glabrous, including the midrib; margin thickened and cartilaginous, not ciliate, bearing a few rather obscure teeth above; petiole 2–3 mm. long, weakly jointed. Inflorescence to 20 cm. long, simple or once-branched below; indumentum as on the stem, but hairs more appressed and mixed with sessile glands and sparse glandular hairs. Calyx 3/2·5 mm. long, turbinate, with appressed hairs and scattered glands; fruiting calyx not seen. Pedicel 3 mm. long. Corolla 8–9 mm. long. Anthers c. 1 mm. Nutlets not seen.

BRAZIL, Bahia: Serra do Sincorá, Nov. 1906, *Ule* 7348 (K!).

Apparently a distinct species, but further collections are needed.

TAB. 3771. *Eriope obovata*. FIG. 1, flowering shoot, × ⅔; 2, leaf, upper view, × 2; 3, calyx, side view, × 6; 4, calyx, inner surface, × 6; 5, calyx, outer surface, × 6; 6, corolla, side view, × 6. (All from *Ule* 7348, holotype.)

(*See page 93*)

18. **Eriope arenaria** *Harley*, sp. nov., *E. crassipedi* affinis, sed habitu suffruticoso, caulibus ramosis tenuibus tenacibus, ramulis sub-basalibus sterilibus copiose instructis differt, setis longis plerumque nullis, foliorum lamina breviore (8–18 mm. longa), angustiore ((1–)2–4(–6) mm. lata), petiolo 1–3(–4) mm. longo obscure articulato etiam distinguenda.

Subshrub, with numerous wiry branching stems 15–30 cm. high, from a woody nodulose rootstock; many slender non-flowering shoots arising from near base of stem. Stems thinly tomentose, with short spreading hairs mixed with glandular hairs in the upper part; long setose hairs absent. Lamina 8–18 mm. long, (1–)2–4(–6) mm. wide, narrowly elliptic to almost linear, coriaceous, with attenuate-acute apex and attenuate base; upper leaf-surface sparsely hairy with scattered broad-based hairs, lower surface subglabrous except for scattered hairs along the midrib and the veins; margin obscurely serrate, more distinctly toothed on broader leaves, weakly ciliate; petiole 1–3(–4) mm. long. Inflorescence 10–20 cm. long, simply racemose, slender; indumentum as on the upper stems with numerous glandular hairs. Calyx 2–2·5/1·5–2 mm. long, turbinate, rather villous and with short-stalked glands, becoming 5–6·5/4–5 mm. long in fruit. Pedicel 1–2 mm., becoming 2–3 mm. long in fruit. Corolla 7–8 mm. long, violet. Anthers c. 1 mm. Nutlets 2–2·5 mm. long.

BRAZIL, Minas Gerais: Serra do Cipó, alt 1100 m., planta de campo arenoso e pedrogoso con flores azuis, 3 Dec. 1949, *A. P. Duarte* 1963 (K!, holotype; RB! isotype); Serro do Cipó, campo, 23 Aug. 1951, *Magalhäes* s.n. (UB!); Serra do Cipó, 16 Jan. 1951, *Kuhlmann & Edmundo* 6 (RB!); Serra do Cipó, 6 km. north of Palacio, nos campos do alto da serra, Oct. 1953, *F. Segada-Vianna & J. Loredo Jr.* 1068 (R!, US!); Serra do Cipó, c. 140 km. N of Belo Horizonte, alt. 1250 m., campo slopes and sandstone outcrops, 19 Feb. 1968, *Irwin, Maxwell & Wasshausen* 20496 (NY!, UB!); Paraopeba, 12 Nov. 1959, *Heringer* s.n. (HB!).

Since its first discovery in 1949, this rather distinctive species has been collected several times, and yet has remained undescribed until now. The type collection was originally identified as *E. filifolia*, another species of the Serra do Cipó, but one which differs markedly. *E. arenaria* is at present known only from the two localities cited above, scarcely 100 km. apart. Related to *E. crassipes*, it differs chiefly in its wiry branched habit, with many non-flowering leafy shoots arising from near the base, and its much smaller, narrower leaves.

TAB. 3772. *Eriope arenaria.* FIG. 1, habit, × ⅔; 2, flower, side view, with part of calyx removed to show corolla, × 8; 3, calyx, side view, × 8; 4, calyx, outer surface, × 8; 5, calyx, inner surface, × 8; 6, leaf, upper view, × 3; 7, leaf, upper view, × 3; 8, portion of leaf, lower surface, × 6. (All from *Duarte* 1963, holotype, except 7 from *Irwin et al.* 20496.)

(*See page 95*)

19. **Eriope xavantium** *Harley*, sp. nov., *E. crassipedi* affinis, sed statura elatiore 1–1·5 m. alta, foliis multo angustioribus glabris breviter petiolatis, inflorescentia haud glandulosa, calycibus longioribus recedit. Ab *E. arenaria* habitu, foliis glabris longioribus differt.

Branched? herb or shrub to 1–1·5 m. tall; lower parts unknown. Stems in the upper parts with very short spreading hairs. Lamina (30–)40–45 mm. long, 4–4·5 mm. wide, narrowly oblong, somewhat coriaceous, with acute apex and narrowly attenuate base; both surfaces glabrous from the first; margin thickened or somewhat recurved, not ciliate, with few, obscure teeth in upper half; petiole 1·5–2 mm., jointed near the base. Inflorescence c. 20 cm. long, unbranched (in only extant specimen); indumentum as on

Eriope crassipes Benth. subsp. *cristalinae* Harley & subsp. *trichopoda* (Briq.) Harley

3771

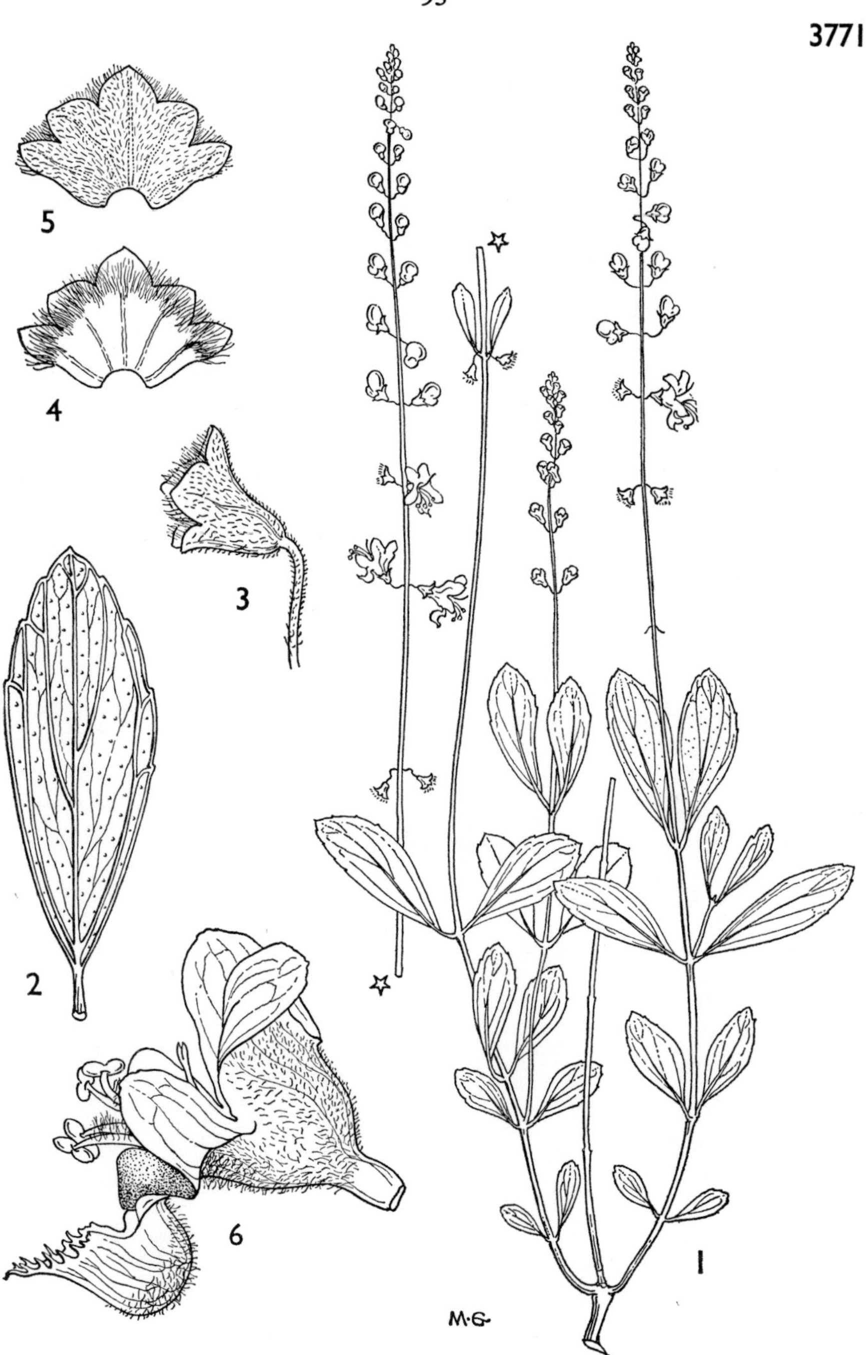

Eriope obovata Epling

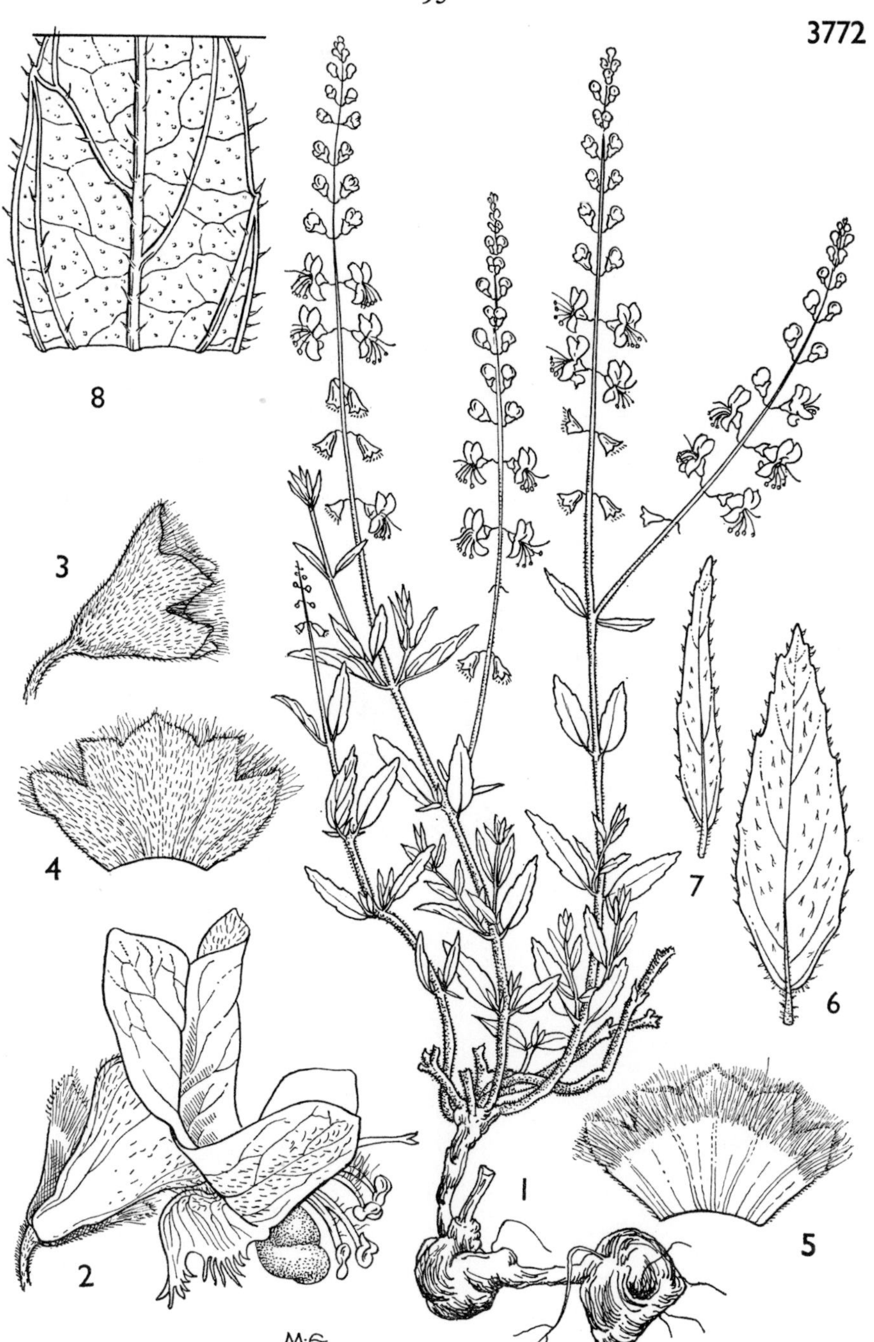

Eriope arenaria Harley

the stems, apparently not glandular. Calyx 3–4/2·5–3 mm. long, turbinate with appressed hairs below, unknown in fruit. Pedicels 1·5 mm. long. Corolla 7·5–8·5 mm. long, 'lilac'. Anthers c. 1 mm., yellow. Nutlets not seen.

BRAZIL, Mato Grosso: 92 km. north of Aragarças on Xavantina road, open sandy area, 14 Jan. 1968, *Philcox & Fereira* 4054 (K!, holotype).

In spite of the paucity of material, this is clearly a distinct species.

TAB. 3773. *Eriope xavantium.* FIG. 1, flowering shoot, × ⅔; 2, calyx, side view, × 6; 3, calyx, outer surface, × 6; 4, calyx, inner surface, × 6; 5, flower, side view, with part of calyx removed to show corolla, × 6; 6, leaf, natural size. (All from *Philcox et al.* 4054, holotype.)

(*See page 99*)

20. **Eriope complicata** *Mart. ex Benth.*, Lab. Gen. et Sp.: 144–5 (1833). Types: Brazil, Bahia, in sylvis catingas et in campis ad Sincorá, *Martius* s.n. (M!, lectotype; UCLA!, US!, photos); Brazil: Minas Gerais, in alta planitie Chapada do Paranan, *Martius* 1831 (M!, paratype).
E. goyazensis Briq., in sched., *synon. nov.* Type: Brazil, Goiás, Chico Costa au bord du bois, *Glaziou* 21825 (G!, K!, P!).
[*E. pallens* sensu Epling, in Fedde Rep. Spec. Nov. Beih. 85 (3): 193–4 (1936), *non* Benth.]
[*E. alpestris* sensu Epling, in *loc. cit.*, non Benth.]

Herb to about c. 50 cm., with several erect usually unbranched stems from a thickened woody base. Stems densely tomentose with short greyish hairs mixed with numerous long spreading setose to villous hairs, the two hair-types becoming more villous and less distinct on the upper part of the stem. Lamina 14–18(–22) mm. long, 10–18(–19) mm. wide, broadly ovate-elliptic to rotund, rugose, usually with a broadly rounded apex and a rounded base; upper surface densely appressed-hairy, lower surface densely villous with long hairs; margin crenate, with numerous teeth; petiole 1·5–3(–4) mm. long. Inflorescence 15–20 cm. long, once-branched below; indumentum as on upper parts of stem, but usually mixed with numerous shortly stalked glands. Calyx 2–3/1·5–2·5 mm. long, turbinate, villous with long somewhat appressed greyish hairs and a few glands, becoming 6–7·5/5–6 mm. long in fruit, narrowly infundibuliform with a broadly rounded spreading upper lip. Pedicel 1·5–2 mm. long, becoming 2·5–3·5 mm. long in fruit. Corolla 4·5–7 mm. long, violet (rarely white) with darker venation. Anthers c. 1 mm. Nutlets c. 3·25 mm. long.

BRAZIL. Goiás: Chico Costa, 10 Oct. 1894, *Glaziou* 21825 (G!, K!, P!); Formosa, 20 Oct. 1961, *Heringer* 10705 (K!, NY!, UB!).
Brasília, Distrito Federal: 1964, *Barroso* 670 (RB!, UB!); 8 Sept. 1965, *Irwin et al.* 8075 (NY!); Sobradinho, 14 Sept. 1964, *Irwin, Soderstrom et al.* 6229 (K!); Chapada da Contagem, c. 20 km. N of Brasília, alt. 1000 m., 5 Sept. 1965, *Irwin et al.* 8004 (K!, UB!).

Bahia: ad Sincorá, November, *Martius* s.n. (M!).
Minas Gerais: Chapada do Paranan, *Martius* 1831 (M!).

In spite of its high degree of morphological uniformity, this plant was misinterpreted by Epling, who confused it with both *E. pallens* and *E. alpestris*, sometimes determining duplicates of the same gathering by different names.

Recent gatherings have been entirely from Gioás and the Federal District of Brasília. I have been unable to locate the whereabouts of the Chapada do Paranan, in Minas Gerais, which may be in the same general area, but the Bahia locality is much more distant, and could conceivably be an error due to the inadvertent switching of labels. Further collecting around Sincorá might clarify this problem.

The type sheet at Munich also includes a small fragment of some leguminous plant, which is obviously not intended.

TAB. 3774. *Eriope complicata*. FIG. 1, flowering shoot, × $\frac{2}{3}$; 2, leaf, lower view, × $1\frac{1}{2}$; 3, portion of lower stem, × 3; 4, calyx, side view, × 8; 5, calyx, inner surface, × 8; 6, calyx, outer surface, × 8; 7, corolla, side view, × 6; 8, fruiting calyx, upper view, × 6; 9, nutlet, dorsal view, × 6. (All from *Irwin et al.* 8004.)

(*See page 101*)

Excluded or Doubtful Names

Eriope elegans Briq. in Bull. Trav. Soc. Bot. Genève 5: 114–5 (1889) = *Hyptis elegans* (Briq.) Briq. ex Mich.

E. *horridula* Epling in Fedde Rep. Spec. Nov. Beih. 85: 191 (1936) = *Lippia* sp. aff. *grandiflora* Mart. & Schau., see Harley in Kew Bull. 28: 121–122 (1973).

E. *micrantha* Benth. in DC. Prod. 12: 141 (1848) = *Hyptis effusa* S. Moore.

E. *nudicaulis* Briq. in Bull. Trav. Soc. Bot. Genève 5: 116–7 (1889) = *Ocimum nudicaule* Benth. Lab. Gen. & Sp.: 14 (1832), *synon. nov.*

E. *trichopes* Epling in Bull. Torr. Bot. Club 71: 495 (1944) = *Hyptis trichopes* (Epling) Harley, in Kew Bull. 28: 24 (1973).

E. *alpestris* Mart. ex Benth. var. '? *glabrior*', Lab. Gen. et Sp.: 145 (1833) ined. The specimen, from Bahia, appears to be an undescribed taxon, but is inadequate.

Collections not Taxonomically Assigned

Irwin et al. 23866, Brazil: Minas Gerais, somewhat viscous subshrub to c. 1·5 m. tall, c. 48 km. W of Montes Claros, road to Agua Boa (950 m.) Serra do Espinhaço, 25 Feb. 1969 (K!).

De Haas Sr., J. H. de Haas & R. P. Belém 219: Brazil: Goiás, Parque Nacional do Tocantins, estrada Veadeiros-Colinas, 8 km. W of Veadeiros. Em campo cerrado, 23 Sept. 1967 (U!).

Both the above may represent new taxa, but further collections are needed to ascertain their status.

3773

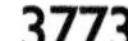

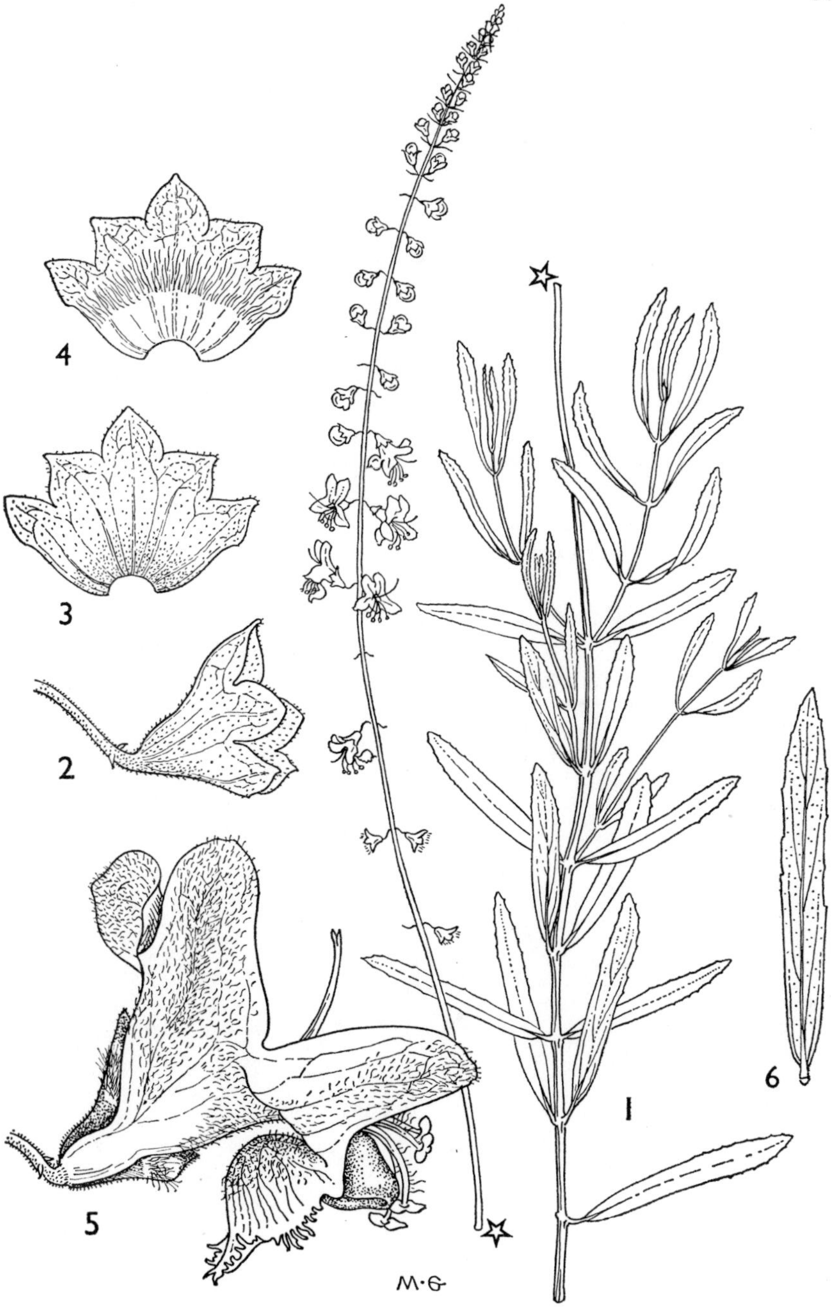

Eriope xavantium Harley

3774

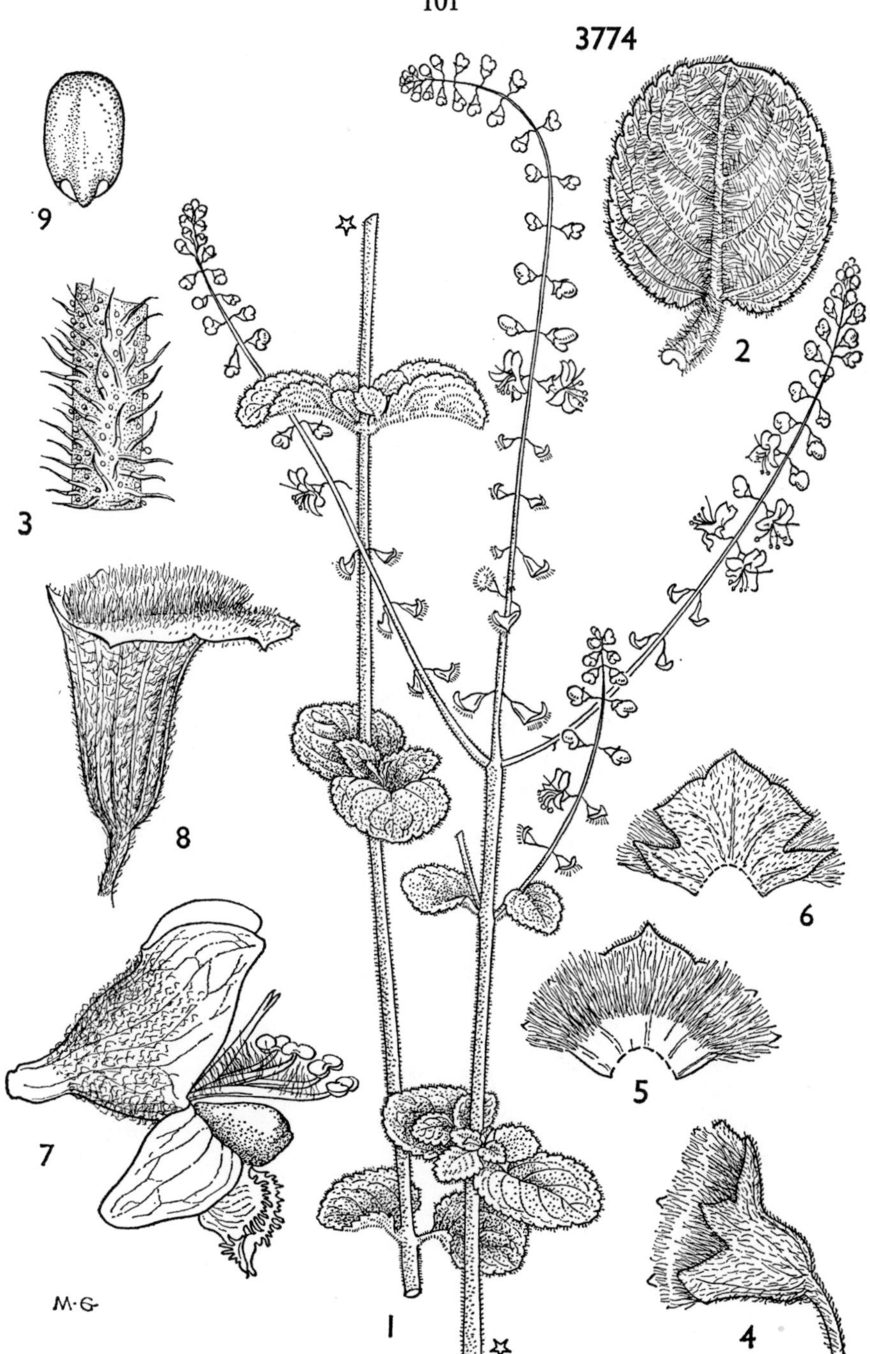

M·G

Eriope complicata Mart. ex Benth.

Eriopidion

Eriopidion *Harley*, gen. nov., ab affini *Eriope* bracteis pedicellos subtendentibus persistentibus, calyce faucem versus angustato, dente dorsali hygroscopico statu sicco deflexo et faucem occludente, statu madefacto patente. Calyx in fauce serie pilorum robustorum ornatus. Corollae tubus angustus, elongatus, supra ipsam basin constrictus, dorsaliter saccatus. Stylopodium nullum. Nuculi trigoni.

E. strictum (*Benth.*) *Harley* comb. nov. Type: Brazil, Piauí, common in dry rock and sandy places near Oeiras, *Gardner* 2279 (K!, holotype; NY!, UCLA!, US!, photos; BM!, G!, K!, OXF!, P!, UCLA!, W!, isotypes).
Eriope stricta Benth. in DC. Prod. 12: 142 (1848) [sphalm. *striata*].

Herb with single erect stems 25–60 cm. long, shortly pubescent with crisped hairs and long spreading setae in the lower parts, becoming more densely crisped-hairy above with some glandular hairs. Lamina (15–)38–60 mm. long, (9–)12–20 mm. wide, narrowly, rarely broadly, elliptic, slightly rugose, with a narrow rounded apex and attenuate to a narrowly cuneate base; upper surface thinly pubescent when young, becoming very sparsely so, lower surface densely hairy, more or less incanous, rarely both surfaces with scattered villous hairs, denser beneath; margin rather bluntly serrulate with numerous teeth; petiole (5–)12–16 mm. long. Inflorescence 28–40 cm. long, with few simple elongate branches, with dense crisped hairs mixed with glandular hairs. Calyx 2·5/2 mm. long, turbinate, with short broadly triangular teeth, dorsal tooth very broad, with appressed hairs; calyx becoming 7/5 mm. long in fruit, narrowly campanulate, narrowed towards the mouth, and with ten straight ribs which anastomose just below it, two-lipped, upper lip distinctly three-toothed, the two outer teeth small, the inner much larger, broader than long, folded when dry to close the calyx mouth, spreading when moist, lower lip indistinctly two-toothed, calyx-throat with a line of coarse hairs. Bracteoles rather broadly lanceolate, scarcely subulate as in *Eriope* species. Pedicel c. 1 mm. long, deflexed in fruit, not enlarging, subtended by a persistent bract. Corolla 5–5·5(–7) mm. long, with narrow elongate tube constricted just above base and saccate near base on dorsal surface so as to appear bent, colour unknown. Anthers less than 1 mm. Stylopodium absent. Nutlets 3–3·25 mm. long, narrowly elongate, triquetrous with convex outer face, smooth.

BRAZIL, Piauí: near Oeiras, Mar. 1839, *Gardner* 2279 (BM!, G!, K!, OXF!, P!, UCLA!, W!); Oeiras, (1878), *Jobert & Schwacke* 1027 (R!).
Ceará: sine loc., *Saldanha* 8249 (R!).
Maranhão: Municipio de São Raimundo das Mangabeiras: c. 45 km. due NNE of Balsas, or c. 45 km. WSW of São Raimundo das Mangabeiras, on the road between these two cities c. 7°10′ S, 45°50′ W, cerrado of scattered trees, grassy ground-cover with herbs, alt. c. 300 m., 15 Mar. 1962, *G. & L. T. Eiten 3672* (US!).

A somewhat isolated species showing several unique characters. These include the persistent bracts, the broad bracteoles, the narrowly campanulate fruiting calyx with hygroscopic dorsal lobe, the narrowly elongate corolla-tube, absence of stylopodium and the triquetrous nutlets. It would seem to be derived from the less specialised *Eriope*, and I now place it in a separate, monotypic genus.

In the herbarium of the Museu Nacional, Rio de Janeiro, there is a *Glaziou* specimen, No. 19685a, clearly identifiable as *E. strictum*, purporting to come from 'Cipó ao Rio Preto, Minas Gerais', almost 1000 miles from the type locality. Wurdack, in Taxon 19(6): 911–13 (1970), has demonstrated that Glaziou distributed other collectors' specimens, including those of Schwacke, under his own name and number, and with false data. In view of the extraordinary discrepancy in the locality, and the striking similarity in condition between the specimens of Jobert & Schwacke and that of Glaziou, I have no hesitation in assuming the latter to be a duplicate of *Jobert & Schwacke* 1027, originally collected in Piauí.

Various Gardner sheets in European herbaria from Piauí are numbered 2278 and 2274, but these are not present at Kew. One such sheet in Vienna has been changed from 2274 to 2294. It seems highly probable that these numbers are due to error, and that all such sheets represent isotype collections.

TAB. 3775. *Eriopidion strictum*. FIG. 1, flowering shoot, × $\frac{2}{3}$; 2, calyx, side view, × 10; 3, calyx, inner surface, × 10; 4, corolla, side view, × 8; 5, gynoecium, × 14; 6, calyx when dry, lower view, × 6; 7, calyx when wet, lower view, × 6; 8, nutlet, ventral view, × 8; 9, nutlet dorsal view, × 8; 10, nutlet, *t.s.*, × 8. (All from *Gardner* 2279, holotype.)

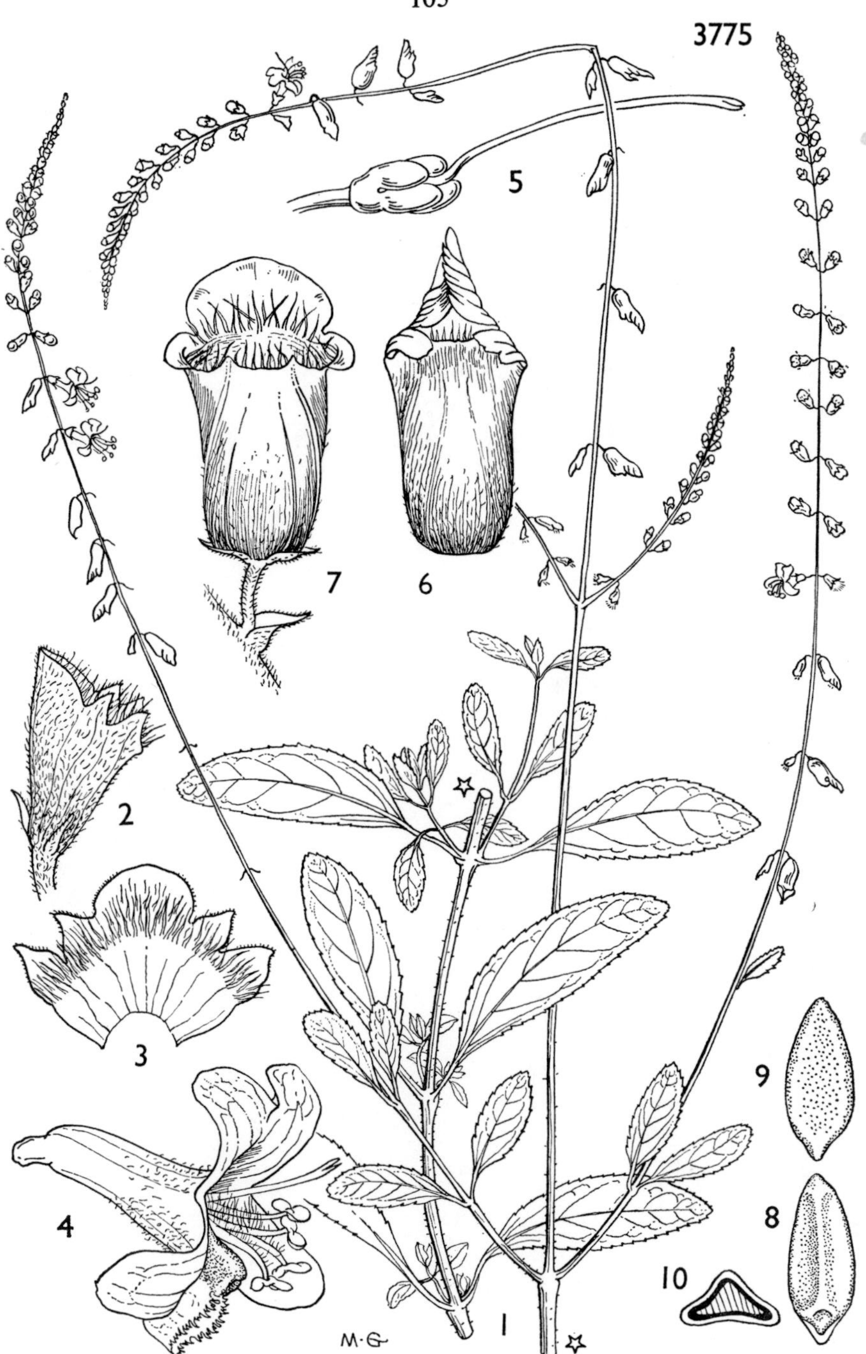

Eriopidium strictum (Benth). Harley

INDEX OF NAMES

(Numbers as in numbering of species in text. Synonyms are italicised and followed by = and
the number of the taxon under which they are included).